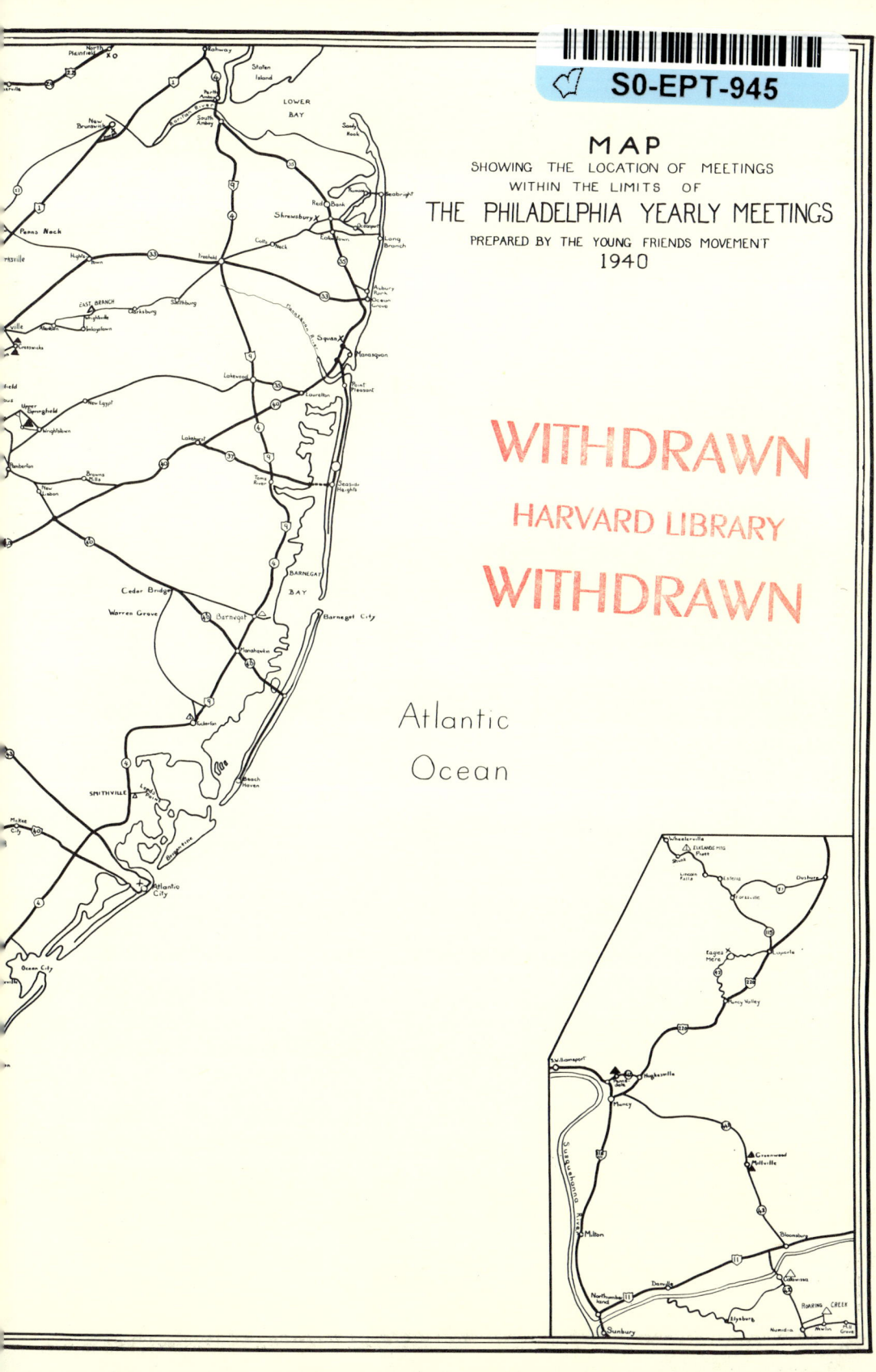

*Friends in the Delaware Valley:
Philadelphia Yearly Meeting 1681–1981*

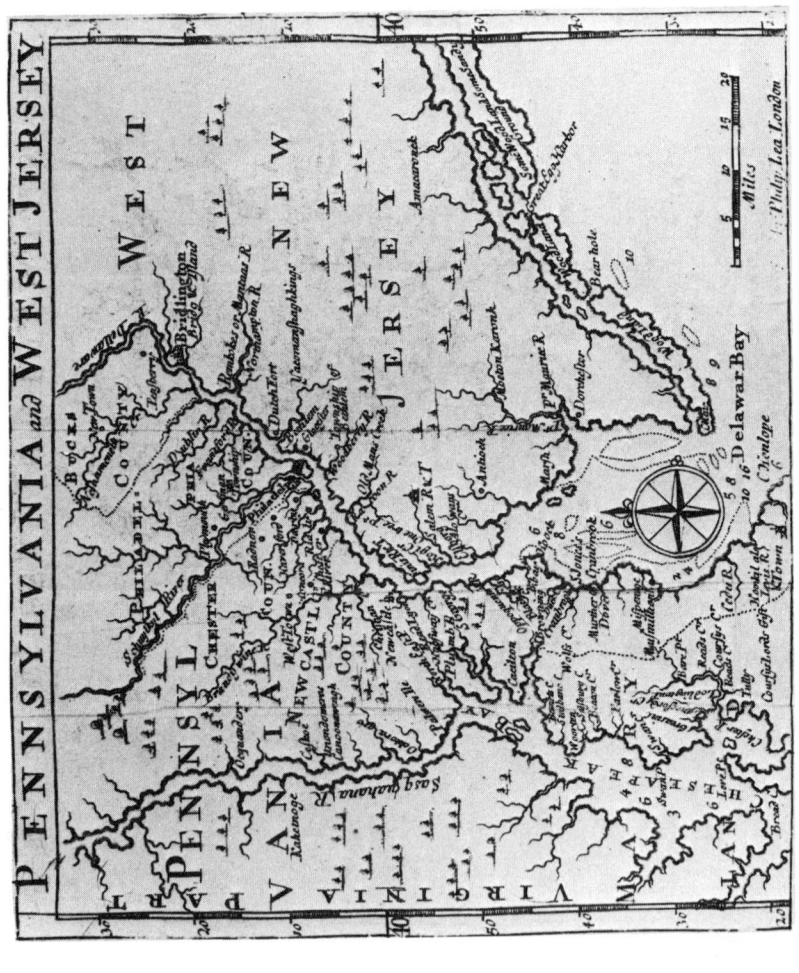

From Gabriel Thomas's *An Account of Pensilvania and West New Jersey,* 1698.

Friends in the Delaware Valley: Philadelphia Yearly Meeting 1681–1981

edited by
John M. Moore

1981

FRIENDS HISTORICAL ASSOCIATION • *Haverford, Pennsylvania*

BX
7607
.P4
F74

The cover design is a late eighteenth century drawing of the Hexagonal Meeting House (*c.* 1685–1785) at Burlington, New Jersey. Philadelphia Yearly Meeting met there in even numbered years until 1760. Artist unknown. Reproduction courtesy of The Quaker Collection, Haverford College Library.

Copyright © 1981 by Friends Historical Association
All Rights Reserved
Library of Congress Catalog Card No. 81-66931
Printed in the United States of America by Dorrance & Company

This book is dedicated
to the memory of
Henry J. Cadbury
Frederick B. Tolles
Eminent Quaker Historians

Contributors

Margaret Hope Bacon, Assistant Secretary for Information, American Friends Service Committee. Her most recent books are *Valiant Friend: The Life of Lucretia Mott* and *As The Way Opens: The Story of Quaker Women in America.*

Edwin B. Bronner, Professor of History and Librarian, Haverford College. He is the author of *"The Other Branch:" London Yearly Meeting and the Hicksites,* and the editor of *An English View of American Quakerism, 1877,* and *American Quakers Today,* as well as other books.

Barbara L. Curtis has recently retired as Quaker Bibliographer, Haverford College Library.

J. William Frost is Howard M. and Charles F. Jenkins Professor of Quaker History and Research, Swarthmore College, and Director of the Friends Historical Library. He is the author of *The Quaker Family in Colonial America* and other books.

Herbert M. Hadley has recently retired as Executive Secretary of the Friends World Committee for Consultation, Section of the Americas, and is now working on a history of that organization.

Mary Hoxie Jones is Research Associate in Quaker Studies in the Quaker Collection, Haverford College Library. She is the author of *Swords into Ploughshares, Quaker Poets Past and Present,* and other books.

Arthur J. Mekeel is editor of *Quaker History* and author of *The Relation of the Quakers to the American Revolution.*

Milton Ream is a member of the Associated Committee of Friends on Indian Affairs.

Elizabeth Gray Vining is the author of many books, including *Windows for the Crown Prince, Being Seventy,* and *Friend of Life: The Biography of Rufus M. Jones,* which has recently been reprinted.

John M. Moore is Professor Emeritus of Philosophy and Religion, Swarthmore College, and President of the Friends Historical Association.

Contents

Introduction *John M. Moore*		1
I	The Founding Years, 1681–1789 *Arthur J. Mekeel*	14
II	Years of Crisis and Separation: Philadelphia Yearly Meeting, 1790–1860 *J. William Frost*	57
III	A Time of Change: Philadelphia Yearly Meeting, 1861–1914 *Edwin B. Bronner*	103
IV	Diminishing Separation: Philadelphia Yearly Meetings Reunite, 1915–1955 *Herbert M. Hadley*	138
V	A Widening Path: Women in Philadelphia Yearly Meeting Move Toward Equality, 1681–1929 *Margaret Hope Bacon*	173
VI-A	Philadelphia Friends and the Indians *Milton Ream*	200
VI-B	The Yearly Meeting and Japan *Elizabeth Gray Vining*	215
VI-C	Philadelphia Yearly Meeting and the American Friends Service Committee *Mary Hoxie Jones*	234
	Appendices *Barbara L. Curtis*	248
	Index	263

Introduction

by JOHN M. MOORE

This book is a series of essays by recognized experts and specialists who deal with crucial periods and significant aspects of the history of Philadelphia Yearly Meeting of the Religious Society of Friends. While it does not profess to be the complete or definitive story, for this would require several volumes, it makes a major contribution to such a history. Most of the essays follow a chronological pattern and are intended to portray the more important events and developments of successive periods from the establishment of the Yearly Meeting in 1681 through the separations and near separations of the nineteenth century to the reunion of the two then existing Yearly Meetings in 1955. Although occasional references are made to events and developments since 1955, we have not attempted a detailed narrative of the last twenty-five years. It is probably too early for such an attempt to meet with much success, for the historian needs perspective in order to distinguish between ephemeral events and those which have lasting significance. A few issues are summarized at the close of this Introduction.

The opening essay, contributed by Arthur J. Mekeel, is based primarily upon a careful re-reading of the minutes of the Yearly Meeting from its establishment in 1681 to the close of the American Revolution and the establishment of the new federal government under the Constitution in 1789. This chapter portrays the growth of Quakerism in the Delaware Valley, first on the eastern or New Jersey side of the Delaware River and then only a few years later on the western side in Pennsylvania and the three lower counties which eventually became Delaware. This growth was extremely rapid; by 1789 Philadelphia Yearly Meeting comprised ten quarterly meetings and at least one hundred particular or preparative meetings—figures which are not far removed from those of today. But the growth in Quaker population was far less than that of the population as a whole, and the Quakers found themselves in a minority early in the eighteenth century. Nevertheless they maintained political control in Pennsylvania until 1756 and a strong influence until the Revolutionary War. The primary emphasis of the chapter, however, is on internal growth and the development of organization and discipline. As problems arose, advices and testimonies were formulated to meet them. The attentive reader will notice a steady development in these matters; sometimes we are surprised by the rapidity of change while at other times we are amazed by its apparent slowness. But changes are continually being made; Quakerism had not yet become a tradition to be preserved and defended. With modifications in the rules of discipline came also differences in organization and ways of securing compliance. In those days Quakers were not content to formulate and uphold ideals; they believed in laying down rules of conduct to which all Friends were expected to conform. This was particularly true after the reform movement of the 1750's had taken hold. This movement was a reaction to what was perceived as a growing laxity in enforcing the rules of discipline, and it resulted in a marked tightening not merely of the rules themselves but in the strictness with which they were applied and enforced. Elders and Overseers exercised increasing power as rules became more specific and enforcement more uniform and strict.

These developments in Quaker discipline were assisted by the growing efficiency of the Society's organization. By the close of the eighteenth century Philadelphia Yearly Meeting possessed a hierarchy of local, monthly, and quarterly meetings, each subordinate to the next higher body as well as to the Yearly Meeting itself. Furthermore, the Yearly Meeting through the Meeting of Ministers and Elders and the more recently formed Meeting for Sufferings possessed highly effective organs of control through which uniformity of discipline could effectively be maintained. Thus the Society of Friends came through the crisis of the French and Indian War and the even greater crisis of the Revolution with its organizational struc-

Introduction

ture intact and probably strengthened. Its success in persuading members to free their slaves provided evidence of sensitivity to a great social problem and proof of the effectiveness of organization and discipline.

In one respect, however, the Society of Friends lagged behind other denominations. It remained unaffected by the formation of the Federal Union and Friends did not form a national body, as for example, the Episcopalians and the Presbyterians did. The six yearly meetings on the east coast remained separate and distinct, and as Friends moved westward across the Appalachians in large numbers after the Revolution they established new yearly meetings in Ohio, Indiana and beyond. Nevertheless, the yearly meetings preserved essential unity among themselves and with the parent body in London through traveling ministers and the exchange of epistles and other correspondence.

The beginning of the nineteenth century saw Philadelphia Yearly Meeting united and apparently prospering. The yearly meeting extended over the widest range of territory in its history and its membership probably reached the highest peak at this time. When the great westward migration began Philadelphia Yearly Meeting remained largely unaffected.

As J. William Frost points out in the second chapter, no one could have foreseen at the beginning of the nineteenth century that the most disastrous schism in its history would soon divide not only Philadelphia Yearly Meeting but most other yearly meetings in America. Much has been written about the great Separation of 1827–28 and many theories have been advanced to explain it. Frost provides an admirably succinct and penetrating account of most of these theories and of the Separation itself. But the puzzling question remains, why did the eighteenth century synthesis, as Frost aptly calls it, break down? The Quaker form of this synthesis was largely the work of Robert Barclay and William Penn who systematized the more original but less coherent ideas of George Fox and the First Publishers. We see this synthesis most perfectly expressed in the writings of John Woolman, whose thinking is both Biblical and rational, conservative and radical. He assumes what William Penn asserted, that Quakerism is completely in accord with primitive Christianity and with the insights of right reason and Pure Wisdom.

But this eighteenth century Quaker synthesis broke down as the nineteenth century advanced for reasons which are partly intellectual and partly historical and sociological. The various strands which had formed the synthesis now tended to unravel, and Friends began to see one another or even to identify themselves as quietists, evangelicals, or liberals; that is, as adherents of a party or faction rather than simply as members of a comprehensive community. Nor were Friends alone in so doing. Similar things were happening in other religious groups. The Congregationalists were dividing into Unitarians and Trinitarians, the Presbyterians into New School and Old School. The Quaker schism was similar to these but not identical with either. In a sense it was a three-way division rather than simply a two-party split, and when schism finally occurred, after more than a decade of increasingly vigorous controversy, it did not strictly follow ideological lines. Each of the two antagonistic bodies which emerged from the separation believed itself to be Philadelphia Yearly Meeting and continued to use the name. For convenience we call these bodies by their nicknames, Orthodox and Hicksite, designations which once provoked resentment but which no longer have invidious connotations.

Each of these meetings soon found itself divided by internal strains and factional stresses. The Orthodox body comprised a quietist or conservative majority and an evangelical minority. While managing to avoid an actual split (except for the defection of a small number who called themselves Primitive Friends) during the controversies of the 1840's and 1850's, it did so at the price of cutting itself off from all official correspondence with other yearly meetings. The division within the Hicksite Yearly Meeting was perhaps not so deep or obvious but it was nonetheless real. It was between conservatives who maintained traditional Quaker attitudes of quietism and isolation from worldly concerns, and liberals who favored cooperation with others in attacking such evils as intemperance and slavery. The conservative element controlled the Yearly Meeting during most of the nineteenth century. The liberals mostly acquiesced, though a small group in the 1850's formed the Yearly Meeting of Progressive Friends at Longwood.

The changes which gradually took place in both yearly meetings between 1860 and 1914 are described by Edwin B. Bronner in Chapter III. At any particular time the differences between the two

Introduction

Quaker bodies were considerable and probably seemed even greater to those within either fold than to an outsider. For a long time both bodies held themselves aloof from the outside world and from each other. The Arch Street (Orthodox) Yearly Meeting was so concerned to avoid a separation between the Wilburite and Gurneyite factions that it ceased to exchange epistles and other official correspondence with other yearly meetings. The Race Street Yearly Meeting exchanged epistles and correspondence with other Hicksite bodies but remained isolated from London Yearly Meeting and from the Orthodox yearly meetings in America. But as Edwin Bronner points out, the situation changed as the nineteenth century wore on, gradually at first and then more rapidly after the turn of the century. Both yearly meetings were on parallel courses and moving in the same direction. Though the pace might be different the movement was similar in both groups: toward less aloofness from the outside world and greater concern with social problems, toward more liberal attitudes in matters of theology and religious belief, and toward greater willingness to recognize one another as Friends and to cooperate in matters of common concern.

By the turn of the century the more progressive forces were coming into their own in both yearly meetings and were increasingly able to move their respective bodies in new directions. Among Orthodox Friends this first took the form of voluntary associations organized by the Gurneyite or progressive party to promote various concerns or causes. In the early 1900's some of these organizations were given official recognition and absorbed into the structure of the Yearly Meeting. Among Hicksite Friends the procedure was often similar but recognition by the Yearly Meeting usually came more quickly. The organization of the Friends General Conference in 1900 was a clear victory for the liberal or progressive element not only in the Race Street Yearly Meeting but among Hicksite Friends generally.

Before coming to the movement toward greater unity which is described in Chapter IV, it seems advisable to call attention to Chapters V and VI which deal with special topics. The first of these is Margaret Bacon's chapter on the women's Yearly Meeting and the changing role of women in the Society of Friends. Margaret Bacon points out that interest and concern in these matters go back to the earliest days of the Quaker movement, to George Fox and Margaret

Friends in the Delaware Valley

Fell. Quaker women enjoyed spiritual equality and were recognized as ministers from the beginning. George Fox organized women's meetings in order to increase their opportunities for service and for expressing their particular interests and abilities. The system of separate meetings of men and women for business and discipline worked well for many years and contributed greatly to the sense of autonomy, responsibility, and self-worth which Quaker women enjoyed. But in the latter years of the nineteenth and the early years of the twentieth century Quaker women became increasingly conscious of the limitations to their autonomy and equality which were involved in the system of separate meetings. Matters of common concern could be dealt with only in a very cumbersome way, and decisions reached in the women's meeting often led to nothing unless approved by the men. As a result sentiment in favor of abolishing the separate women's meetings began to develop and reached a similar conclusion in both the Arch Street and Race Street Yearly Meetings during the 1920's. It is noteworthy that the first clerk of the combined Race Street Yearly Meeting in 1924 was a woman, Jane P. Rushmore, a worthy successor to the eminent position held in the women's Yearly Meeting a generation earlier by Lucretia Mott.

The final chapter in the book is a composite one. It deals with the general subject of Quaker outreach, but in three specific fields: Philadelphia Friends and the Indians, Philadelphia Friends and Japan, and the contribution of Philadelphia Friends to the establishment and the continuing activities of the American Friends Service Committee.

Friends had to face the question how to deal with the Indians from the beginning of their settlement in the Delaware Valley. Indeed William Penn had sent advance emissaries to purchase land from the Indians, to apprise them of the impending settlements, and to promise fair dealing and friendly relationships with the Quaker settlers. William Penn was remarkably wise and extraordinarily successful in his dealing with the Indians, and the Quakers were the beneficiaries of his wise and generous policies for many years thereafter. But after Penn's death, his sons succeeded to the proprietorship. They did not remain within the Quaker fold nor did they maintain consistently their father's policy of fair dealing and friendly relationships. The Quakers did not seek to monopolize settlement in Pennsylvania but

Introduction

opened the new colony freely to others, and the Quakers were soon outnumbered. Many of the newer arrivals settled on the frontier and came into contact—often hostile—with the Indians. The old policy of fair dealing and friendly relationships was replaced by mutual suspicion and hostility. When actual conflict came in the French and Indian War the Quakers formed a voluntary organization, the Friendly Association, and worked to restore harmony. After the Revolution the Yearly Meeting established an Indian Committee and sent representatives to work among the Senecas and other tribes to bring about reconciliation and to help the Indians to become farmers and adapt to new conditions. It is not necessary here to rehearse the whole story of Quaker involvement with Indians and Indian problems. Milton Ream tells the story succinctly and well. Here we need only remark that the problems are with us still.

Elizabeth Gray Vining has a delightful chapter on Friends in Japan which she is uniquely well qualified to write. The work in Japan began as a fairly conventional missionary activity and was supported by a voluntary organization formed by the Gurneyite wing among the Orthodox Friends. These Friends were closest in spirit and attitude to evangelical American Protestantism in the nineteenth century and were therefore most affected by the missionary ardor which prevailed among most religious bodies during that period. The Quaker missionaries responded more quickly than most to changing concepts and emphases and the Japan Mission, though small, became a recognized model for effective work and willingness to share responsibility with the Japanese themselves. The ultimate outcome, in spite of the immense difficulties created by war and occupation, is a small but effective Japan Yearly Meeting and a large and highly regarded school for girls, the Friends Girls School in Tokyo.

Mary Hoxie Jones is on familiar ground when she describes the contributions of Philadelphia Friends to the organization of the American Friends Service Committee and to its continuing activities. Her chapter is focused not on the establishment and work of the AFSC—this story has been told more than once though never completely—but primarily on the ways in which various groups and committees of Philadelphia Quakers have contributed to and assisted in this work. Hence details of her chapter will be unfamiliar to many present-day readers, and it is highly desirable that the whole story should be placed in the record. Furthermore, the whole enterprise was immensely significant and valuable for Philadelphia Quakerism itself, and over the space of a generation (1914-1955) helped immeasurably to bring about the reunion of the two Yearly Meetings.

This theme is more fully treated in Herbert Hadley's chapter which deals specifically with the movement toward reunification of a divided Quakerism during this period.

Hadley tells how various Quaker groups and organizations gradually—sometimes hesitatingly—learned to work together on common concerns, to cooperate in various ways, and eventually to merge. Then he describes how a similar process was carried out by several committees of the two Yearly Meetings, and how the Yearly Meetings themselves first learned how to cooperate, to hold joint sessions, then to form the General Meeting, and eventually to reunite. It is a long and involved story, though never a tedious one, for the eventual goal was kept clearly in mind and rarely deviated from. If any of us, looking back, is tempted to exclaim, "How long, how long it took those Quakers to come together," we should remind ourselves that when reunion finally came it came with great relief and cordiality on both sides. Only rarely have the mergers of religious bodies been carried through so successfully and completely.

One perhaps unexpected result of the reunion has been a greater participation by Philadelphia Friends in the ecumenical movement. The Arch Street Yearly Meeting had taken a more active role in this movement than the other and was a member of the National Council of Churches as well as of the corresponding state and local bodies. Many Race Street Friends were active in local ecumenical programs and through the Friends General Conference were affiliated with the World Council of Churches. After the reunion there was widespread willingness to participate in ecumenical endeavors at all levels, and these relationships were maintained by the united yearly meeting. This attitude of openness and cooperation contrasts strongly with the aloofness and isolation which Friends had maintained toward other churches in earlier periods.

The focusing of attention so completely on the movement toward reunification of the two Yearly Meetings during the period from 1914 to 1955 means that other significant developments have received little attention. Some of these developments have indeed been touched

Introduction

upon. For example, the first World Conference of Friends held in London in 1920 and the second held at Swarthmore and Haverford in 1937, with the consequent establishment of the Friends World Committee for Consultation, have been described in some detail. But no mention has been made of the participation of Philadelphia Friends in the third World Conference at Oxford in 1952 or in the fourth at Greensboro, North Carolina in 1967. The elimination of separate Yearly Meetings for Women in the 1920's has been described, but nothing has been said about the gradual decline or change of status of the Meeting of Ministers and Elders which also has taken place during the last half-century.

When the Race Street Yearly Meeting amended its Discipline in 1918 the provisions for recording ministers and appointing elders were eliminated and the Meeting of Ministers and Elders was replaced by a Committee on Ministry and Counsel. The object of these changes apparently was to open up the spoken ministry more widely to Friends and to democratize what was commonly referred to as the "Select Meeting." The Arch Street Yearly Meeting never adopted a similar change, but in many monthly meetings the practice of recording ministers simply fell into disuse. The new book of Discipline, adopted at the time the Yearly Meetings were reunited, permitted monthly meetings to continue the practice of recording ministers if desired, although the actual practice had virtually ceased. The Meeting of Ministers and Elders was replaced by a Committee on Worship and Ministry in the monthly meetings and a Meeting on Worship and Ministry at the Yearly Meeting level. The Meeting on Worship and Ministry was intended for all members of Committees on Worship and Ministry in monthly meetings but was open to any Friend who wished to attend. For a few years this Meeting was presided over by its own Clerk, the Clerk of the Yearly Meeting stepping aside for this session. But this practice was soon discontinued and the Meeting on Worship and Ministry has become a Committee of the Yearly Meeting and reports to the Yearly Meeting like any other commitee. It is probably too early to assess the ultimate effect of these changes. Some Friends regard them as merely changes in terminology, but others believe that they indicate significant changes in fundamental attitudes. To be appointed a member of a committee on Worship and Ministry for a few years is much like being appointed to any other committee; it is hardly the same as being recognized as a minister or elder under Divine appointment.

Other changes in the character and structure of our meetings have come about so gradually that they have scarcely been noticed but probably have considerable significance. One is the virtual disappearance of preparative meetings; another is the apparent decline in importance and significant function of quarterly meetings. In earlier times, and in fact through most of the nineteenth century, monthly meetings were composed of two or more preparative meetings, sometimes as many as five or six. Chester Monthly Meeting, for example, was composed of four preparative meetings before the separation, and Concord of six. A similar situation prevailed throughout the Yearly Meeting. The separation resulted in many changes in details but the general pattern remained the same in both Yearly Meetings. But as the nineteenth century progressed many preparative meetings sought to become monthly meetings and came to be recognized as such by their quarterly meetings. The reasons for this change and the factors which brought it about have not been thoroughly studied, but one may surmise that desire for greater autonomy in the local meeting was an important factor. Nowadays the local or particular meetings are nearly all monthly meetings and almost all preparative meetings have disappeared.

Some comments should also be made about the changing status of quarterly meetings. It is hard to make valid generalizations on this point because the present quarterly meetings vary enormously in strength and vitality. Some are quite strong while others have declined greatly. But all of them have been affected by pervasive changes and have lost functions which were formerly important. One traditional function of quarterly meetings was to hear and adjudicate appeals from disciplinary decisions of monthly meetings. But disciplinary cases have become so rare as to be virtually non-existent, and hence there is nothing in this area for the quarterly meeting to do. Quarterly meetings also appear to be losing the function of appointing representatives to Yearly Meeting and to its committees. Most of these appointments today are made either by monthly meetings directly or by the Yearly Meeting Nominating Committee. The quarterly meetings still perform important financial functions and many Friends believe that they provide a valuable link between monthly

Introduction

meetings and the Yearly Meeting. It would be rash therefore to predict that quarterly meetings are destined to disappear; it is enough for the historian to say that we appear to be in a period of transition and that no one is sure where we will come out.

Another change of recent years which should at least be mentioned is the steady decline in membership reported by the Yearly Meeting during the last twenty years. In 1960 the Yearly Meeting reported a total membership of approximately 17,600 of whom 12,600 were adults and 5,000 were children (minors). In 1980 the total was 13,758 (11,540 adults and 2,218 children). The decrease in adult membership over this period is almost exactly one thousand, and many would say that this is the result of monthly meetings pruning their membership lists of inactive members. The decrease in minors has been much greater; the present figure is less than 50 percent of that reported twenty years ago. It is sometimes said that the decrease in number of children is more apparent than real, since parents are less inclined than formerly to register their children as members. But since attendance of children at First Day Schools has also shown a sharp drop, it is hard to resist the conclusion that the decrease in membership is serious and portends ill for the future unless the trend can somehow be reversed.

During the same period the Yearly Meeting has increased the amount raised through the quota from $101,000 to $457,000. This expansion has been made possible by increasing the quota or assessment per adult member from $7.91 in 1960 to $39.60 in 1980, by the modest success of the Combined Appeal and by a considerable increase in the use of income from bequests to meet current budget needs. Like other organizations the Yearly Meeting has found it increasingly difficult to keep up with inflation. Those responsible for the budget and finances warn that a crisis is imminent.

When the editor looked over the following chapters for the last time before sending them to the printer, he became convinced that insufficient attention has been given to intellectual factors, at least at certain points. Activities and accomplishments have been emphasized more than beliefs and doctrines. More attention has been given to social testimonies and their practical applications than to changes in the intellectual content and climate of Quakerism. There are, of course, significant exceptions to such generalizations. J. William

Friends in the Delaware Valley

Frost has contended that the separation of 1827 was produced in part because what he calls the eighteenth century consensus was breaking up. We ought, I think, also to point out that reunion became possible in the mid-twentieth century because a new consensus developed. The beginnings of this may be seen in the 1890's and it developed rapidly after the turn of the century. Rigid theological dogmas gave way to more liberal philosophical attitudes. Biblical criticism lent support to the view that religious beliefs are formed in a process of cultural evolution and may be expected to change and develop over time. The view that religious experience is more important and more fundamental than the changing categories in which it is expressed intellectually became increasingly popular. Rufus Jones's theory of the mystical nature and origin of Quakerism came to be generally accepted in both branches of the Society and proved to be an extremely powerful influence for dissolving differences. Thus a new and more liberal outlook came to prevail, and a greater tolerance for remaining differences made possible the gradual achievement of unity.

The new found unity of Friends in Philadelphia Yearly Meeting has been subjected to many strains and stresses during the quarter-century that has passed since the reunion in 1955. The mystical theory of the origins of Quakerism has been vigorously challenged by those who hold that the Quaker movement is to be understood primarily as an outgrowth of radical or sectarian Puritanism. There has been a revival of interest in theology and a renewed willingness to discuss issues which remain controversial. The liberal consensus of the last generation seems to be gradually weakening, but what will take its place is by no means clear. The theological differences which still exist among Friends in this Yearly Meeting do not appear to endanger unity, since there is now a widespread willingness to accept such differences as normal or even to value them as signs of vitality and as pointing the way to larger conceptions and more inclusive truths.

More serious perhaps are our differences in social and economic philosophy and in political allegiances. The Yearly Meeting appears to be composed of a moderate majority and a radical minority which is vigorous and articulate. These differences give rise to sharp controversy in various committees, particularly those dealing with mat-

Introduction

ters of Quaker testimonies and social concerns, and sometimes in sessions of the Yearly Meeting. Friends will long remember the controversies which resulted from the demands of the Black Economic Development Conference and those which occurred during the Vietnam War, particularly when the decision was made to send medical supplies to both North and South Vietnam. The Yearly Meeting showed remarkable resiliency in dealing with these crises, and it is a matter of deep satisfaction that it came through them with increasing strength and confidence.

I The Founding Years, 1681-1789

by ARTHUR J. MEKEEL

When Phineas Pemberton was appointed clerk to the Yearly Meeting in 1696, one of his tasks was to collate all previous minutes of the Yearly Meeting, heretofore kept in an indiscriminate manner, and institute an official minute book. As a preface to the minutes he inscribed the following "Epistle":

> It hath pleased God in his infinite goodness and good providence to give us his people who are in scorn called Quakers a lot and inheritance in this new and remote and formerly to us unknown part of the world now called America, into which desert and wilderness he hath called, drawn and allured many of us and hath given us of the comfort of his house and abundantly blessed us by pouring down of his mercies upon us, both inwardly and outwardly...[1]

Four decades earlier in 1658 three Quakers headed by Josiah Coale on their way from the Chesapeake to Boston were the first Friends to refer to the Delaware Valley region. The first Quakers to settle within the confines of what was to become Philadelphia Yearly Meeting moved from Long Island to Middletown and Shrewsbury in East New Jersey in 1665 and established Shrewsbury Monthly Meeting in 1670. When George Fox was traveling among Friends in America in 1672, he visited these Friends as he passed back and forth from the Eastern Shore to Long Island.

Two years later, in March 1674, Lord Berkeley sold West New Jer-

---

1. Philadelphia Yearly Meeting, Minutes, v.1, p.1. Quaker Collection, Haverford College. Hereafter cited as PYM, Min.

The Founding Years

sey to two Friends, John Fenwick and Edward Byllynge. It was not long before Quaker settlements began to appear along the Delaware and in each case a meeting for worship was established almost immediately. In March 1681 William Penn received the magnificent grant of the province of Pennsylvania, and Friends in West Jersey knew that they would soon be joined by thousands of their fellow religionists who would settle opposite them on the other side of the river.

A few months after the grant of Pennsylvania to William Penn took place in England the first "general or yearly meeting for East and West Jersey" was held on August 31, 1681, at the home of Thomas Gardner in Burlington on the eastern shore of the Delaware. Friends had settled at Salem in 1675 and in the area around Burlington a few years later. Now representatives from the monthly meetings at Salem (1676), Burlington (1678), and Chesterfield (1680) were joined by Friends from Marcus Hook and Upland (Chester) (1681), and Falls west of the Delaware to establish what was to become the largest and most influential Yearly Meeting in the new world.[2]

In the course of that day's meeting, this small group of Friends laid the basis for the rapid development of the Yearly Meeting in the succeeding years. They agreed that women's meetings should be held at the same time as men's meetings, that all meetings for worship should begin at ten o'clock, that Friends who desired to "travel in the service of truth" should obtain the concurrence of their respective monthly meetings, and that a general meeting for worship should be held annually at Salem on the second First Day of the Second Month (April in the old calendar).[3]

The rudiments of the discipline are seen in the provision that Friends should refer their differences to the monthly meeting rather than go to the law and that each monthly meeting should appoint two persons (later called overseers) to find out who reported what was said in meeting and who circulated false reports so that the

---

2. At the request of this meeting, Friends on Long Island and Rhode Island (New England Yearly Meeting) agreed to transfer Shrewsbury Monthly Meeting to the new Yearly Meeting in New Jersey in 1682 where it remained throughout the colonial period. Aside from Chester, the monthly meetings west of the Delaware were organized subsequent to 1681. PYM, Min., v.1, pp. 2,3.
3. Until 1752 the calendar year began in March (the Julian calendar). The names of months in this essay have been changed to correspond with modern usage.

offenders might "be dealt with according to the merits of the offense." The following year the Yearly Meeting directed that any persons contemplating marriage should obtain certificates of good conduct from their respective meetings before the marriage was allowed.

In an epistle of advice to the constituent meetings, a device subsequently regularized, the Yearly Meeting of 1682 exhorted Friends not to wear "superfluity of apparel" nor indulge in immodest or unseemly use of tobacco, nor sell useless things, but that they should rather "keep within the bounds of moderation and the Fruits of the Spirit of Truth."

Before the next Yearly Meeting in 1683, William Penn had arrived and established his new colony. Over the next decades there flowed a tide of Quaker immigrants from England, Wales, and Ireland, and within a few years there were three quarterly meetings in Pennsylvania, namely, Philadelphia, Chester, and Bucks, in addition to three in New Jersey: Burlington, Salem, and Shrewsbury. At the same time increasing numbers of non-Quaker immigrants arrived in the new colony of Pennsylvania. By 1700 the Friends were already outnumbered by an expanding worldly society with its concomitant impact on the Quaker way of life.

At the next General Meeting held at the home of Thomas Brit in Burlington on September 4, 1683, William Penn and other Friends recently arrived from Britain were present. The meeting appointed a committee including Penn to propose to Friends in the other colonies the establishment of a continental yearly meeting and also to write to the Friends in England "to give an account of the affairs of truth here."[4] The annual exchange of letters with London Yearly Meeting, which seems to have actually begun a year later, played an important role in creating a strong bond between the two yearly meetings.

With the large number of Friends now settled in Pennsylvania, General or Yearly Meetings began to be held on both sides of the

---

4. The idea of a continental yearly meeting, probably proposed by the newly arrived British Friends who were not yet aware of the geographical constrictions of the American colonies, was prompted by the British practice of sending representatives from the quarterly meeting in each county and Wales to London Yearly Meeting. (PYM, Misc. Pap., 1662-1702, p. 9.) Friends from Chop Tank and Herring Creek Quarterly Meetings in Maryland and Rhode Island in New England were present at the Yearly Meeting of 1686 but they made no commitment as to a continental yearly meeting. PYM, Min., v.1, p.9.

The Founding Years

Delaware with some resulting confusion. Therefore it was decided in 1685 that there should be but one Yearly or General Meeting held alternately in Philadelphia and Burlington which would begin on the first Sunday in September, the first three days to be devoted to public meetings for worship and the fourth day to the men's and women's business meetings. The meeting further decided that Friends in the Ministry should meet on First Day at seven o'clock in the morning preceding the public meeting for worship.

The latter action led to the establishment of the Meeting of Public Friends or Ministers, subsequently the Meeting for Ministers and Elders. It delegated to itself the function of approving all publications by "those who profess the faith in these American parts." Its meetings were to be held three times a year: on the first Saturday in June, the Saturday before the Yearly Meeting, and the first Saturday in March.[5] It was to this body that traveling ministers reported. The accounts of these journeys as recorded in its minutes indicate the importance of these widespread and frequent contacts with Friends from the Carolinas to New England in fostering cohesion within the Society.

Over the next few years the organizational structure of the Yearly Meeting was largely completed. The quarterly meetings were directed to send two representatives each to the Yearly Meeting "capable to give an account of the affairs of truth." In 1693 the representatives were directed to sit as a body before the yearly meeting sessions "to communicate what business they find amongst themselves and others to be of necessity." On a few occasions representatives were present from the lower counties of New Castle, Sussex, and Kent. Although authorized to set up their own quarterly meeting, Friends in these counties evidently chose to be a part of Chester Quarterly Meeting. In 1695 the Yearly Meeting established a central fund or yearly meeting stock to carry on-running expenses. Each quarter was assessed a proportionate annual payment to maintain this fund.[6]

During these years the Friends were greatly concerned to maintain the good relations with the Indians initiated by William Penn. In

---
5. Eventually two such meetings were held, the general spring meeting and the annual meeting at yearly meeting time.
6. From 1688 on the term "General" was not used in designating the yearly meeting.

1685 the Yearly Meeting informed the subordinate meetings that the sale of rum to the Indians was inconsistent with the "honor of truth," an admonition repeated the following year. However, the lucrative nature of this practice so tempted some Friends that, in its epistle to the subordinate meetings in 1687, the Yearly Meeting declared that abstention from "selling rum or other strong liquors to the Indians directly or indirectly or exchanging rum or other strong liquors for any goods or merchandise with them" was to be thereafter considered a testimony. It further directed that all Friends in their respective monthly meetings should subscribe personally to this testimony.

In the sphere of personal conduct the Yearly Meeting warned Friends against evil speaking and backbiting under penalty of disownment. Friends were also cautioned against the unseemly use of tobacco and were urged to observe due regard "in taking of it both as to time and place that our Holy Profession be not reproached thereby." Finally, Friends were to desist from wearing the hat in time of prayer and were to avoid controversy over the matter.

One problem which came before the Yearly Meeting at this time and which was increasingly to engage and prick the conscience of Friends was slavery. In 1688 there came before the Yearly Meeting a protest against the holding of slaves initiated by Germantown Friends and endorsed by the Monthly Meeting at Dublin (Abington). Although the Yearly Meeting declined taking decisive action at that time on the matter, "it having so general a relation to many other parts," the protest proved to be the beginning of a long struggle to persuade the Society of the inconsistency of Friends holding slaves.

An important and constant activity of the Society in England from its earliest days was the printing and disseminating of statements on Quaker beliefs, sermons of Quaker leaders, and tracts expounding and justifying the Quaker viewpoint on religious matters. In order to assure that the members of the Society be well informed on religious questions, the Yearly Meeting concluded a contract in 1684 with William Bradford to be the yearly meeting printer, with the provision that all books on Quaker subjects have official approval. The Yearly Meeting also obtained books from England for distribution among the monthly meetings. Subsequently a press was obtained from

The Founding Years

England, the management of which was placed in the hands of Philadelphia Monthly Meeting.[7]

After a decade of peaceful development, the Quaker communities on the shores of the Delaware were shocked by sharp charges of unsoundness leveled against the Society and some of its leading ministers by George Keith, one of their foremost members. Keith, a graduate of Aberdeen University, influenced by Henry More and the Cambridge Platonists, had left his Calvinist faith for the more spiritually inward expression of Christianity which he had found in the early Quaker movement. A diligent co-worker with Fox, Penn, and Barclay, Keith was an able speaker and writer and authored some thirty books and tracts in defense of Quakerism.

In 1685 Keith emigrated to East Jersey as Surveyor General of the province and helped to fix the boundary between East and West Jersey. In 1689 he became master of the new Quaker school in Philadelphia, later the William Penn Charter School. Coming from a dogmatic religious background with a highly structured church organization, he became increasingly disturbed by what he considered the loose theology and discipline of the Friends in frontier Pennsylvania and New Jersey. As a Public Friend or minister, he spent much time visiting meetings in those and neighboring provinces in an effort to strengthen the Society in those areas he found deficient. Finally, in 1690, he presented a paper entitled "Gospel Order Improved" to the Meeting of Ministers, calling for the appointment of deacons and elders in all meetings and for a confession of faith to which all would subscribe.

After extended consideration, the Meeting of Ministers was unable to accede to Keith's proposals as presented, although it offered to accept a confession of faith composed of statements by Christ and the Apostles and to submit the matter to London Yearly Meeting for its judgment. Keith rejected these offers and his acerbic temperament then came to the fore. A quarrel between Keith and another minister, William Stockdale, over soundness in doctrine, which the Meeting of Ministers unsuccessfully attempted to mediate, led to open contro-

---

7. In 1709 a committee of overseers of the press was appointed to judge the acceptability of all manuscripts to be published on the authority of the Yearly Meeting. After 1756 this function was transferred to the Meeting for Sufferings.

versy. Keith proceeded to attack the Society, declaring that "there was not more damnable Errors of Doctrine of Divels amongst any of the Protestant Professions, than was amongst the Quakers."[8] Among other accusations he charged that the Friends considered the Light of Christ Within to be sufficient for salvation and that they largely ignored the outward role and offices of the physical, historical Christ. Even though this charge was denied by Friends, he continued his attack along this line. In the course of the controversy Keith also called for a closer conformity to Quaker testimonies and declared that participation in the exercise of governmental functions must inevitably involve the use of force and was therefore inconsistent with the peace testimony.

Over a period of months Keith and his followers, some of them formerly prominent Friends, established a separate body which they called Christian Quakers. They subjected the Society to increasingly virulent vituperation not only publicly but also in Quaker meetings for business. After lengthy and patient dealings with Keith, the Meeting of Ministers, of which he had formerly been a leading member, issued a testimony against him on account of his "rage and violence against us" which had rendered the Friends "a scorn to the profane" and because of the fact that he had set up separatist meetings. The Yearly Meeting soon followed suit with a testimony of disownment. Both meetings submitted full accounts of the affair to British Friends who subsequently, with the support of Penn, urged further attempts at reconciliation, but with no success. Keith appealed to London Yearly Meeting, but this body, in 1695, disowned him also.

Considering the vitriolic and unrestrained manner in which he acted, it seems obvious that Keith was emotionally unbalanced because of the extreme tensions to which he was subjected. Moreover, one cannot but regret the loss of the contribution Keith might well have made to the development of the Society had he been less adamant in his demands and more conciliatory in his attitude and conduct.

Keith and his followers established fifteen separatist meetings in Philadelphia and Burlington and in Bucks County, calling themselves Christian Quakers. Their discipline provided for deacons and elders

---

8. PYM Ministers and Elders, Min. v.1, p.14.

The Founding Years

and a confession of faith to which all members must subscribe. To their credit they issued a strong indictment of slave holding, a position to which the Society came over a half century later. They were soon torn with dissension, and Keith himself subsequently joined the Anglican Church and was ordained a priest. After an extended visit to England he returned to Pennsylvania where for the next few years he was a thorn in the side of Friends. Some of his followers ultimately returned to the Society while others became Baptists or joined the German Pietists or the Anglican Church. In their epistle to London Yearly Meeting in 1700, the American Friends stated that the Keithians were "almost moldered to nothing."

Another occurrence which caused concern to the Yearly Meeting was the attempt of the annual general meeting for worship at Salem, set up in 1681, to establish itself as a meeting for business. In 1695 it requested assistance from the Yearly Meeting in organizing its business procedures. In order to block the development of a rival body, the Yearly Meeting informed the Salem Friends that the only business which could be transacted at the time of the annual worship sessions was business of the Quarterly Meeting. The Yearly Meeting was prepared to furnish assistance for that purpose alone.

An ever present concern of the Yearly Meeting was the faithful adherence by its members to the peculiar Quaker way of life. From time to time it issued admonitions urging Friends to a consistent observance of their testimonies. In 1689 a committee of nine headed by George Keith sent a statement to the quarterly meetings strongly advising the "Keeping of a Godly and Christian Discipline, and more especially a Tender Inspection over Youth." This was followed two years later by the issuance of a "Testimony against loose and unclean Spirits." At the Yearly Meeting of 1694 Philadelphia Quarterly Meeting submitted a proposal regarding the education of youth, which resulted in the publication of "A General Testimony Against all looseness and vanity," exhorting all Friends to follow the instructions and good order established by Friends in England, and urging the monthly meetings to admonish and labor with those who walked disorderly. It especially warned youth against unseemly sports such as "challenging each other to run races, wrestling, laying of wagers, . . . (and) drinking to one another." Declaring it was a grief to many to hear children of Friends "use the world's language; as you to a single person and amongst one another," the statement exhorted parents to set a good example in their families and not to give their sons and daughters "in marriage with unbelievers." After issuing extended instructions on proper procedures in marriage, it directed that the testimony be read in all quarterly and monthly meetings.

The following year the Yearly Meeting issued strict advices on simplicity in clothing and in house furnishings.[9] Friends in business were exhorted to "keep to a word in their dealings . . . (and not) trade by land or sea beyond their abilities." Each particular meeting was instructed to appoint two or more Friends, men and women, to deal with those who did not heed these advices.

At the same time the Meeting of Ministers, concerned about the quality of the ministry since there were some ministers who were using "unseemly noises, tones and gestures, and drawing their words at a great length with Ahs, . . . going from supplications to exhortations . . . with unsound expressions and imperfect quotations of places in scripture . . .," agreed that there should be meetings of men and women ministers weekly in Philadelphia and quarterly in the country to watch over the ministry.

At the Yearly Meeting of 1696 the question of slavery again arose with greater urgency than before. Several papers were read on the subject, which prompted an advice to be issued urging Friends to "be careful not to encourage the bringing in of any more Negroes." Those who held slaves were exhorted to bring them to meeting or to hold meetings with them and their families.

A year later, in 1697, the first relief assistance project was undertaken. That year there was an acute shortage of food and provisions in the northeastern towns of Massachusetts, and the New England Friends appealed for aid. In response the Yearly Meeting called for the raising of a fund for the general relief of Friends and others in distress in their sister colony. Contributions for this purpose subsequently amounted to £383.

In the first decades after the founding of the colony, the legislative side of the government, which was entirely in the hands of Quakers,

---

9. Among other things Friends were cautioned against wearing long lapped sleeves or long curled periwigs. The appointments indicated were a continuation of similar action taken in 1681 and may well mark the permanent institution of overseers. (See PYM, Min., v.1, pp. 53–55.)

The Founding Years

soon reflected the diverse interests of the country and city Friends. The former, especially affected by the proprietor's land policies, tended to be anti-proprietary in politics. The city Friends, on the other hand, many of them merchants and tradesmen who later constituted the monied aristocracy, were strong supporters of the proprietary interests. The Welsh Friends, who had settled the so-called Welsh Tract where they dreamed of establishing a linguistic and cultural entity of their own, were an important element in the country party, and their leader, David Lloyd, headed the populist country faction in the legislature. As attorney general of the province he espoused the interests of the colonists against imperial and proprietary interests. When Penn returned to America in 1699, he dismissed Lloyd from his post at the urging of the British government, thereby incurring much popular resentment which soon led to bitter political altercations. The conflict between the divergent parties was strongly in evidence at the time Penn granted the Charter of Privileges in 1701.

Although most of the important Friends in the Yearly Meeting held various positions in the government, both legislative and administrative, the Yearly Meeting itself played no direct role in political and governmental activities. Its chief concern was that Friends in the government abide consistently by their testimonies and that they not participate in any action, personal or official, contrary thereto. Therefore, when the conflict of internal governmental factions became obvious, the Yearly Meeting addressed an epistle of advice to the subordinate meetings in 1701 in which it remonstrated sharply that some Friends had conducted themselves in a factious manner in the government, participating in disturbances and parties for personal ends and vengeance. The meeting declared its abhorrence that "any should sacrifice the peace of a province to a private revenge" and directed the monthly meetings to deal with such offenders.

At the conclusion of the epistle the Yearly Meeting stated its position on the relation of the Quakers to government. It declared that for many years the Friends had proven themselves "peaceable subjects of those whom God by his Providence hath set over us, First to the King as supreme, and next to those in authority under Him . . . rendering all their dues."

The final step in the formulation of yearly meeting organization and procedures was the issuing of the first Book of Discipline in 1704. This Discipline, the work of a committee appointed the preceding year to collate all advices and statements on discipline and the testimonies previously issued by the Yearly Meeting, consisted of two parts. Part I, entitled "A General Testimony against all Looseness and Vanity," stressed modesty and sobriety and the avoidance of the vain customs of the world in speech and dress, of "immodest and indecent smoking of tobacco" and of "excess in drinking." Other sections repeated admonitions as to plainness in house furnishings, honesty in business, the training of children under Friends' auspices, and the proper manner for settling personal disputes.

Part II prescribed the organization of the Society and the procedures to be followed by the various meetings and those who were chosen to administer the Society's affairs. Monthly meetings were directed to appoint "two sober and judicious men and two women Friends" to oversee the life of each preparative meeting. It also regularized the appointment of representatives, two men and two women, to attend the quarterly from the monthly, and the yearly from the quarterly meetings respectively. Any Friends who contravened the code of conduct prescribed in the Discipline and persisted therein contrary to the advices of the overseers were to be disowned. Thus, by the opening years of the new century, the Yearly Meeting as well as the colony had established sound foundations on which to face the problems of growth and development over the next half century.

The Middle Years—1705-1754

During the half century following Penn's final departure for England, Pennsylvania expanded rapidly both in population and prosperity. Philadelphia became the busiest port in the colonies, and the province became the wealthiest and one of the most populous. A constant flow of immigration, especially from Germany and Ireland, carried the tide of settlement well toward the Appalachian ranges. In 1733 the Yearly Meeting's annual epistle to the British Friends referred to the "great increase of people, not only of those born among us but others of diverse nations, customs and manners, which of late years have flown in upon us."

The Founding Years

Meantime the Quaker proportion of the inhabitants became ever smaller, and by 1720 the Friends were already a religious minority. Nevertheless they remained the dominant political power, and they prospered economically quite beyond their numerical relation to the rest of the population. From the standpoint of sheer numbers, however, the Society expanded steadily during these years and became more dispersed as the general population spread westward.

Already in 1733 Menallen Indulged Meeting near Gettysburg was set up, and the first minute of a nearby meeting in Huntington states that there were "divers families of Friends of late settled on the west side of the Susquehanna." From there the migration of Friends turned southward through the Shenandoah Valley, and in 1735 Hopewell Monthly Meeting was set up in Virginia. It was not until the revolutionary period that Quakers settled west of the Alleghenies, moving up from meetings in northwest Virginia.

Another wing of migrating Friends moved northwestward into Berks County. By 1718 a Friends meeting was held in Oley near Reading which led to the establishment of Exeter Monthly Meeting in 1737. By 1753 an indulged meeting was set up to the north in Pottsville, and finally in 1775 Catawissa Indulged Meeting was established just south of the Susquehanna River's east branch.

Whereas in 1710 there were sixteen monthly meetings, ten in Pennsylvania and six in New Jersey, with a total of sixty-five meetings for worship, in 1755 there were thirty-one monthly meetings and more than one hundred preparative meetings. Because of the migration to the west, Western Quarterly Meeting was set off from Chester Quarterly Meeting in 1758 to include the more westerly meetings in Pennsylvania and the meetings in northern Virginia.

Although the Yearly Meeting's structure was well established when the Discipline of 1704 was issued, some significant alterations were made in the succeeding decades. In 1712 it was decided to hold a two day public meeting for worship annually in the spring, at first in May but soon changed to the third Sunday in March. Accordingly the Meeting for Ministers designated the Saturday preceding the meetings for worship as the time for its annual spring meeting. Subsequently it became the custom at that time to discuss and determine important matters to be brought before the next Yearly Meeting.

Two years later, in 1714, the Yearly Meeting finally acceded to a request of the Meeting of Ministers and directed the quarterly and monthly meetings to appoint two or more Friends, among them "prudent solid women Friends . . . as well as men" to sit with the meetings of ministers and assist them in caring for the religious life of the respective meetings. Four years later the elders were authorized to sit with the Yearly Meeting of Ministers which thereby became the Yearly Meeting of Ministers and Elders.

It was during this period that regular correspondence with the other American yearly meetings became well established, a practice which with the traveling ministry did much to nurture and maintain the common bonds of the Society. Within a few years of the founding of the Yearly Meeting several ministers visited Friends from New England to the Carolinas. The exchange of epistles with other American Friends was for some time quite sporadic. Not until 1713 did a regular correspondence with New England Friends begin, and not until 1716 with New York and Maryland Friends. Only in 1720 and 1728 respectively did Virginia and North Carolina yearly meetings begin an annual correspondence with Philadelphia Yearly Meeting.

London Yearly Meeting played a special role in the life of the Yearly Meeting. The annual epistles to the Friends in England, which were initiated soon after Penn's arrival in America, took the form of a report on the spiritual condition of the Society, an account of the Yearly Meeting's achievements and plans, and an occasional request for assistance in matters pertaining to the colony's relations with the proprietors or the British government. Subsequently it was the Yearly Meeting through which appeals were channeled to the British Friends to intercede with government officials at times of political crisis. In 1735 the Yearly Meeting appealed to both the London Second Day Morning Meeting and the Meeting for Sufferings to make representations to the British government against a petition of Lord Baltimore to the King in Council to confirm for him a grant of the lower counties. Again in 1740 the Yearly Meeting solicited the aid of the British Friends in countering complaints to the King about the influence of Friends' principles on the actions of the Pennsylvania government.

Contacts with the British Friends were further promoted by the frequent intervisitation of British and American Friends, ministers, and others. Up to the Revolution over one hundred British Quaker

The Founding Years

ministers visited America, most of them spending some time within the environs of the Yearly Meeting.

An ever recurring concern of the Yearly Meeting was that Friends adhere consistently to their peculiar testimonies as required by their "Holy Profession." Apprehension over an increasingly worldly society concomitant with the rapid increase of a non-Quaker population was voiced in the annual epistle of 1731 to the London Friends: ". . . with grief we observe vice and immorality also to increase." It was to guard against such worldly influences that the Ancient Testimony of 1704 was revised and reissued in 1722.

Even more important was the revised Book of Discipline issued in 1719 which elaborated and clarified the accepted code of Quaker conduct and prescribed the means to obtain its proper observance. As a further step to enhance Friends' mindfulness of the standard of life to which they were committed, the Yearly Meeting formulated a set of queries in 1743 to be read in all preparative and monthly meetings at least once in each quarter.

Although the Discipline stressed simplicity and moderation in speech, dress, and home furnishings, the Yearly Meeting never enjoined actual asceticism. In advices issued in 1734 it rather cautioned Friends against immoderate pursuit of worldly riches which if "brought within due bounds for the comfortable subsistence of ourselves and families and the charitable relief of others is not only lawful but commendable."[10]

A recurring problem in the earlier years was the erection of tombstones. The Yearly Meeting repeatedly admonished against such action as well as against the wearing of distinguishing mourning apparel. The excessive use of liquor and tobacco also received attention, and in 1724 the Yearly Meeting directed that Friends who overindulged in strong drink should be "early admonished and dealt with as disorderly persons." It cautioned the aged who partook in pursuit of warmth to beware of "the false heat it seems to supply. . . [since] the true warmth of nature becomes thereby supplanted."

The problem of strong drink was especially acute at weddings and funerals, and the Yearly Meeting declared that "many sober minded among us think needless" the amount of strong liquor and wine

---
10. PYM, Min., v.1, pp. 380-381.

served on such occasions with resulting excesses "grievous to all sober and well minded." It therefore urged that weddings be held on weekdays to "prevent great expectations." The overseers were directed to prevent the unnecessary use of strong drink at burials.

Since commerce and trade were such important aspects of the life of many Friends, the Yearly Meeting early pointed out the implications of the Quakers' religious profession for the conduct of their business. In 1713 it urged Friends engaged in trade to maintain the quality and true weight of goods sold or transported. The Ancient Testimony of 1722 exhorted them to maintain a "religious observance of contracts, bargains and promises," and two decades later, in 1746, the Yearly Meeting reminded its members that the "primitive Friends were remarkable for their uprightness and honesty in commerce and converse." Friends were urged not to become involved in business activities beyond their ability to manage and thereby be a reproach to truth.

Because of the active commercial life of many Friends, legal problems involving court action frequently arose. According to the Discipline all disputes between Friends were to be settled under the auspices of the meeting. Such a procedure, however, was not adequate to deal with types of cases which frequently occurred. Therefore the Yearly Meeting decided that where both parties requested it, and the meeting deemed resort to the courts appropriate, a judicial proceeding was allowable.

An area in which the impact of the world on the society was especially evident was the marriage of Friends to non-Friends, a practice which the Society increasingly believed would weaken the proper observance of the Quaker code of conduct. The Discipline of 1704 strongly admonished against such intermarriage, and in 1712 the Yearly Meeting clearly stated that Friends who married out of the meeting should be dealt with, and if they would not condemn their action, disowned. This admonition was further strengthened in 1722 when immediate disownment was prescribed for those Friends who married non-Friends despite the Meeting's previous advice against such action.

Family life and the nurture of children in the faith received the frequent attention of the Yearly Meeting which considered the young as a vital factor in the future strength of the Society. In letters of

The Founding Years

advice to the subordinate meetings, parents were urged to rear their children "in all plainness both of speech and apparel" and to "use their endeavours to keep their children out of vain, evil and loose company." Instead of allowing "romances, playbooks and other vain and idle pamphlets in their houses and families" they should encourage their children in the reading of the Holy Scriptures and other good and religious books." As a means of avoiding the worldly company associated with various pastimes, parents were advised to keep their children from such unwholesome activities as lotteries, music, and dancing.

In 1729 the Yearly Meeting broadened its concern for the proper rearing of Quaker children and declared that they should be given instruction at least in reading and writing and that then they should be put out to some useful employment, if at all possible in the hands of Friends. Formal education was sporadic and centered largely in Philadelphia. Some monthly meetings had established elementary schools soon after the founding of the colony, but the Friends Public School in Philadelphia, later the William Penn Charter School, accommodated both Friends and non-Friends. As time went on, the Society came to realize the need for a systematic program of basic education for Quaker youth devoid of the undesirable influence of non-Friends. This concern prompted the Yearly Meeting in 1746 to urge the monthly meetings to encourage and assist each other in the settlement and support of schools for the instruction of their children, "at least to read and write and some further use of learning." In establishing such schools the masters and mistresses should be capable of giving not only secular but also religious instruction.

In 1750 the Yearly Meeting repeated this advice and directed the monthly meetings to report progress to the next Yearly Meeting. It further suggested that in order to obtain "religious, prudent persons" as teachers, salaries should be specified as well as the number of children to be taught. Subsequently some monthly meetings complained that the distances in the country where the majority of Friends lived were such that it was difficult to implement the Yearly Meeting's recommendations. Nevertheless, the Yearly Meeting, urged on in its efforts by the British Friends, directed the monthly meetings to continue pursuing the possibilities of establishing such schools, and in 1753 it suggested that the meetings investigate

Friends in the Delaware Valley

whether there was money available for this purpose in charitable funds in their hands. However, it took the emergence of a strong reform movement at the time of the Revolution to bring these initial efforts to an effective conclusion.

The one social and moral issue which became increasingly insistent in its demands on the conscience of Friends during this period was slavery. Already opposition to the importation of Negroes had resulted in legislation by the Quaker Assembly, at first levying duties on imported Negroes and soon thereafter, in 1711, prohibiting the importation of Negroes altogether, an action subsequently vetoed by the crown.

In 1712 agitation against slaveholding prompted the Yearly Meeting to request London Yearly Meeting for a statement of policy on the matter. The London Friends declined taking such action, probably because of the political situation then prevailing in England, although they recommended that Friends should stay out of the African slave trade.[11] In 1715 John Hepburn of New Jersey published a forceful statement against slaveholding, and Chester Quarterly Meeting brought the matter before the Yearly Meeting as it had done in 1711. Fearing disruption within its ranks and lacking support from London Yearly Meeting for decisive action the Yearly Meeting followed a temporizing course. It appealed to its members to "avoid judging one another on this matter publicly or otherwise," but it did direct that Friends who imported Negroes should be dealt with.[12]

The Discipline of 1719 incorporated a section on slavery which cautioned Friends against importing or buying Negro slaves. It also urged those who owned slaves to treat them in a Christian manner and to acquaint them with the principles of truth and morality. Progress in this area was such that in 1728 the annual epistle to London Yearly Meeting claimed that no Friends within the limits of the Yearly Meeting were concerned in importing Negroes.

Two years later, in 1730, three quarterly meetings declared that buying Negro slaves was wrong and should be restricted. In response the Yearly Meeting strongly advised the subordinate meetings that

---

11. London Yearly Meeting chided the Pennsylvania Quakers for not having first consulted with other American Friends so they could present a joint opinion. (PYM, Misc. Pap., 1714, p. 12a.)
12. PYM, Min., v.1, p.168.

The Founding Years

"Friends ought to be very cautious about making any such purchases in the future" and directed that those who did so should be dealt with. Several years later, Benjamin Lay, an active opponent of slaveholding, in order to shock the conscience of the Yearly Meeting, went before the meeting as it was about to break up and squirted blood-colored liquid from a bladder concealed in a book, meanwhile decrying the cruelty of bringing Negroes as slaves from their home countries.[13] Sentiment on the issue had by this time become strong enough that when the queries were first issued in 1743 one query asked: "Do Friends observe the former advices of the Yearly Meeting not to encourage the importation of Negroes nor to buy them after imported?"[14]

Despite action taken to induce Friends not to import or buy slaves, slaveholding as such was widespread in the Society and was even increasing.[15] Quaker slaveholders were among the wealthiest and most powerful men in the colony, were largely merchants and tradesmen in Philadelphia, and were the controlling element in the Society. It was the concern and work of Friends such as Anthony Benezet and John Woolman which gradually aroused the conscience of the body of Friends as to the evils of the slavery system.

In 1754 John Woolman published his *Some Considerations on the Keeping of Negroes,* which had a wide circulation in the Society. It played an important role in finally convincing Friends that their Christian faith required them to issue a testimony against slaveholding, contravention of which would invoke disownment, a measure adopted two decades later. The same year Anthony Benezet penned a statement presented to the Yearly Meeting by Philadelphia Quarterly Meeting condemning the purchase and holding of Negro slaves and calling for official support of this stand with strong disciplinary measures against those who refused to comply.

Such an appeal, coming from that section of the Yearly Meeting where much of the slaveholding was concentrated, indicated the strength of growing antislavery sentiment. As a result the Yearly

---

13. Lay subsequently published *All Slavekeepers, Apostates* without the permission of the Yearly Meeting and was disowned.
14. PYM, Min., v.1, p.434.
15. In 1747 the Swedish traveler Peter Kalm stated that the Quakers held as many slaves as did other people.

Friends in the Delaware Valley

Meeting issued "An Epistle of Caution and Advice Concerning the Buying and Keeping of Slaves." In this epistle it declared it had observed "with sorrow" that slaveholding had increased among Friends and referred the members to its frequently expressed disunity with the "importation and purchasing of Negroes." It charged the overseers to deal with those engaged in such activities and further exhorted Friends to free their slaves in view of the inconsistency of slaveholding with the message of peace and love. However, while enunciating a principle, the Society had not yet reached the point of its full implementation.

As already indicated, most important government positions and many of lesser importance in Pennsylvania were filled by Friends. Leading members of the Yearly Meeting were also leading members of the provincial government. The magistracies were largely filled by Quakers. Despite the decreasing proportion of the Friends to the total population, they were invariably elected to government posts since they were considered as the guardians of the freedom which other people as well as the Quakers enjoyed. Therefore the Quakers exercised complete control of the Assembly until 1756.

Outstanding examples of such prominent Friends were: Phineas Pemberton (d. 1702), clerk of the Yearly Meeting, member of the Assembly, and master of the rolls of the province; Isaac Norris, Sr. (d. 1735), clerk of the Yearly Meeting, mayor of Philadelphia, member at different times of the governor's Council and the Assembly and presiding judge of the court of common pleas; David Lloyd (d. 1731), prominent member of the Yearly Meeting, member and speaker of the Assembly and attorney general of the province; Israel Pemberton, Sr. (d. 1750), clerk of the Yearly Meeting, for nineteen years member of the Assembly; Samuel Carpenter (d. 1714), leading Friend in the Yearly Meeting, member of the governor's Council and reputedly the richest man in the province in 1700; and John Kinsey (d. 1750), clerk of the Yearly Meeting 1730–1750, speaker of the Assembly and chief justice of the province, all positions held co-terminously.[16]

It was thus that although the Yearly Meeting did not involve itself

---

16. Aside from William Penn, no colonial Quaker had the absolute confidence of Friends in church affairs and exercised strong leadership in the state as did Kinsey.

The Founding Years

directly in matters of state, the relationship with the government resembled an interlocking directorate. As already indicated, the chief function of the Yearly Meeting in this connection was to assure that individual Friends serving in government maintained their testimonies faithfully both in their private lives and in the policies they adopted in governmental decisions. Inevitably Quaker principles affected government policies, and the demands of government were bound to create tensions for observant Quaker office holders.

The factionalism in the government evident in 1701 had continued and by 1710 had become even more strident. The Yearly Meeting in the meantime had become increasingly disturbed over this situation. Therefore when the elections of 1710 took place it issued a strong admonition to Friends in the government to be "very careful to act according to Truth and the Testimony of it," and not to "think to excuse a contrary practice by any temptation." It further directed that those who failed to heed this advice and opposed "the judgment of Truth" were to have a "testimony go out against them." Further, the Yearly Meeting reissued the statement of 1701 against factionalism, quarreling and organizing parties.

With regard to the elections, the Yearly Meeting urged Friends to vote for "such of any persuasion who were God fearing . . ., meek, patient, aiming for righteousness and Truth and peace in all things." Furthermore, Isaac Norris, the clerk, published a statement under the authority of the Yearly Meeting appealing to the people at large to "use that great Privilege of the Right of Election, . . . and chuse such men to represent them as have no other view than the true and lasting Good of the Province."[17] This was one time when the Yearly Meeting entered directly into the political field and with the desired result of a complete change in the Assembly membership and the cessation of the previous factionalism.

One of the most sensitive subjects in the field of government was the taking and administering of oaths and affirmations, a matter which affected not only the body of Friends at large but even more those who were magistrates, many of them being ministers. Not until

---

17. *Friendly Advice to the Inhabitants of Pennsylvania* (Philadelphia, 1710, reissue of 1739) pp. i,3. The reissue was made by a non-Quaker, an opponent of Quaker policy, at a time when there was a dispute between the governor and the assembly over the raising of money for defense. John Kinsey was then speaker.

1718 were affirmations accepted as equivalent to oaths by the British government, and that year the Yearly Meeting admonished Friends to keep clear of taking or administering oaths. A fully acceptable form of affirmation, omitting any reference to God, was finally approved by the Crown in 1725, on which occasion the Yearly Meeting submitted an address of thanks to the King.

The action of Quaker magistrates in administering oaths to non-Friends long remained a difficult problem. In 1693 an order from England required all colonial officials to administer oaths to all persons willing to take them, and consequently some Friends resigned their positions in the government. The Discipline of 1719 clearly stated that the overseers should deal with Quaker magistrates or clerks of court who administered oaths, since such action was inconsistent with Friends' principles and was "not any part of their duty as magistrates." In 1733 the Yearly Meeting became even stricter and directed that Quaker justices who allowed oaths to be administered by clerks serving under them were violating this testimony and were to be dealt with. Some Quaker justices subsequently resigned while some continued to administer oaths, primarily in areas where there were few non-Quaker officials, in order to enable the local government to function. Such cases of deviation seem to have often been overlooked.

The Quaker testimony which had the most profound effect on the Friends role in government was the peace testimony. From its origin the Yearly Meeting frequently exhorted Friends to observe this principle in their relations with one another and in the conduct of their meetings. In its advice on conduct in government issued in 1701 the Yearly Meeting pointed out that for many years the Friends had "proven themselves peaceable subjects of the King." The commitment of the Society to pacifist principles was also evident in the refusal of some New Jersey Friends to conform to a militia law passed in 1704, soon after that province became a crown colony. In 1710 Salem and Gloucester Quarterly Meeting submitted to the Yearly Meeting an account of the sufferings of some Friends who refused to bear arms.

It was not until the issuance of the Discipline of 1719, however, that a clear and explicit statement of the testimony was made. The Discipline strongly admonished Friends to be vigilant in maintaining

The Founding Years

their peaceable principles and to avoid joining in any warlike preparations, whether defensive or offensive. Nor were they to purchase prize goods which by their nature were tainted with warlike action. It was the responsibility of the overseers to deal with any who carried guns on their ships for defensive purposes and with any who were concerned in war either on privateers or as owners of ships sailing with guns or with letters of marque. Those Friends who refused to heed such advice were to be disowned.

Among the Friends who held positions in the Pennsylvania government there was some division of opinion as to the role of the pacifist testimony in relation to matters of state. Until the Seven Years War in 1756, the Friends in the Assembly acquiesced at times in a *modus vivendi* in complying with royal requests for financial assistance on the occasion of colonial wars through such devices as granting funds "for the King's use." Thereby they avoided an outright and clear rejection of the Quaker peace testimony. In 1739, when war with Spain loomed on the horizon, the governor presented a request for defense funds to the Assembly. John Kinsey, who was speaker as well as clerk of the Yearly Meeting, spearheaded a drawn out controversy with the governor in opposition to such a grant, and in the end the Quaker position prevailed. The obligation of members of the Society to observe faithfully the requirements of the peace testimony was unequivocal, and that same year the Yearly Meeting drew up a statement exhorting Friends to adhere strictly to their peaceable principles and "in no manner to join with such as may be for making warlike preparations, offensive or defensive."

Despite the firm position taken by the Yearly Meeting, there were some Friends, especially those who were politically active, who questioned the expediency of an absolute application of the peace testimony to defensive warfare. A leader of this group, James Logan, formerly Penn's chief assistant and long a member of the governor's Council, addressed a letter to the Yearly Meeting in 1741 arguing that circumstances were such that the Friends in the Assembly should resign and thereby allow the government to take such defensive action as the crown desired. A committee of the Yearly Meeting, after considering the letter, decided that it should not be presented since it concerned military affairs of the government, a subject inappropriate for the meeting's attention.

Friends in the Delaware Valley

The wider implications of the peace testimony became evident in a decision of the Yearly Meeting in 1742. Chester Quarterly Meeting reported that a militia law had been passed in the lower counties (Delaware) which exempted Friends from certain requirements on presentation of a certificate of membership in a Friends meeting. The Yearly Meeting advised that the granting of such a certificate amounted to complicity in the execution of the law and was therefore contrary to the peace testimony.

The same year a letter from the British Friends urged the American Quakers to stand firm in their "ancient and Christian testimony against wars and fightings and being concerned in any warlike preparations." Two years later they called for the disownment of any Friends who carried guns to defend their ships or were concerned in privateering. After faithfully following this advice and disowning a number of members for such offenses, the Yearly Meeting complained to the Friends in Britain that the problem had been greatly intensified by the bad example set by some British Friends and suggested that the quarterly and monthly meetings in Britain investigate the matter.

By the middle of the century the Anglo-French colonial rivalries were becoming ever more acute, and in 1750 an opponent launched a virulent attack on the pacifist position of the Friends as it affected the policies of the government on frontier defense. John Smith, an intimate friend of John Woolman, composed a clear and effective reply in his *Doctrine of Christianity* which quickly received the approbation of the overseers of the press of the Yearly Meeting and was published in an edition of one thousand copies.

There was never any doubt as to the nature of the Quaker pacifist position, and by this time it had become clear that an ultimate solution to the conflict between political demands and religious conviction could not long be delayed. In fact, the denouement of the conflict was soon brought about by the final determining struggle between the French and British empires in America.

The Years of Crisis and Reform—1755-1789

The first half of the eighteenth century was a period of peace and prosperity for Pennsylvania, and the steady economic growth of the

The Founding Years

province led to the rise of a rich social aristocracy which included leading Quaker families.[18] A trend to laxity on the part of many Friends in observing the standards of simplicity and sobriety, characteristic of the earlier years, prompted the Yearly Meeting as early as 1747 to recommend to the quarterly and monthly meetings that they revive the appointment of "solid and weighty Friends and elders with ministers to visit families" and thereby prevent "many growing inconsistencies and customs amongst us."[19]

The acquisition of great wealth on the part of the Quaker merchant princes of Philadelphia inevitably brought with it a temptation to luxury and ostentatious living clothed in a superficial plainness bordering on hypocrisy. It was this trend that aroused Anthony Benezet, John Woolman, and John Churchman, representing a newer generation of Friends, to a vigorous espousal of a return to a faithful observance of the primitive Quaker testimonies and way of life. There also were those, especially in the political life of the colony, who attempted to combine their Quaker adherence with the practical ways of the world. Isaac Norris II, who became speaker of the Assembly on the death of John Kinsey, exemplified this element in the Society.

Isaac Norris continued the tradition of James Logan in supporting defensive warfare and refused to surrender his Assembly seat, along with other politically minded Quakers, when the Assembly voted for defensive preparations in 1756. As head of the Quaker Party until his resignation in 1764, he was one of the most influential and highly respected men in the province. Although clerk of his monthly meeting and generally observant of the Quaker way of life, he was highly cultured and widely read and did not always conform with what strict Friends desired. He was a leading member of the Quaker aristocracy whose compromise between Quaker witness and active involvement in worldly affairs was rejected by those who called for renouncing the "cultivation of the outer plantation" and a return to the "cultivation of the inner plantation."

At the same time as Isaac Norris II succeeded to the political leadership of John Kinsey, Israel Pemberton, a strong advocate of the

---

18. Such families included the Pembertons, Drinkers, Morrises, Walns, Mifflins, Biddles, Powels, Hills, Fishers, Penningtons, Reynells, Logans, and Coates.
19. PYM, Min., v.2, p.4.

revitalization of the Quaker testimonies and way of life, became clerk of the Yearly Meeting. By 1755 concern over the state of the Society resulted in the appointment of a committee by the Yearly Meeting for the revision and strengthening of the queries. This committee included John Churchman, John Woolman, Samuel Smith, and Samuel Fothergill who was in the American colonies on a visit from England and was a strong supporter of spiritual renewal. The chief changes in the queries were the addition of a strong statement on "bearing arms, training or military service" and a firm directive for the immediate disownment of anyone marrying out of the Society. Laxity in the enforcement of the latter rule came increasingly to be seen as allowing the infiltration of influences weakening strict adherence to Friends' principles.

As a further step in "repairing the walls of Zion and restoring the desert places," the meeting directed the establishment of women's meetings where they had been omitted. The Yearly Meeting of Ministers and Elders meanwhile called for the setting up of meetings of ministers and elders in those monthly and quarterly meetings where there were none, and it also issued a set of queries for ministers and elders concerning their personal conduct and the performance of their spiritual functions.

In order to secure compliance with its decisions, the Yearly Meeting appointed a committee of thirty ministers, elders and other solid Friends to visit the monthly and quarterly meetings and inspect their procedures for maintaining the discipline, and to assist the meetings when needed. Among the members of this committee were John Pemberton, John Woolman, Daniel Stanton, and Samuel Fothergill. The following year the women's meeting was asked to appoint some members to the committee, "with a single eye to the exaltation of Truth and the edification of the Church." This committee became the spearhead for the revitalization of the Society, and it was continued for several years.

The effect of these measures was felt widely throughout the Yearly Meeting and resulted in a strong reemphasis on the exclusion of worldliness from the Quaker way of life. It also laid the basis for a strict observance of the peace testimony and reinvigorated the drive to eliminate the practice of slaveholding by members of the Society.

In order to reinforce these efforts, the Yearly Meeting in 1761

The Founding Years

directed the Meeting for Sufferings to review a collation of minutes adopted from its inception concerning the testimonies and rules of discipline which had been extracted by James Pemberton. The result was the revised Discipline of 1762 which tightened some of the disciplinary procedures, strengthened the queries issued in 1755, and incorporated the latest statements of the Yearly Meeting on slavery and the conduct required of Friends in government.

In succeeding years the reform activities of the monthly meetings resulted in a far greater incidence of disownments than in the previous decades. Between 1760 and 1775 disownments for failure to observe the testimonies probably affected twenty-two percent of the membership of the Society as it stood in 1760. In Philadelphia Quarter sixteen percent of the membership of 1760 incurred this penalty.

When measures for the revitalization of the Society were initiated, the consistent maintenance of the peace testimony was a paramount issue. By this time French and British frontier rivalry had instigated Indian attacks on settlers, and various colonies took steps to improve their defenses. Friends in the non-Quaker colonies of New Jersey and Delaware were soon affected, and early in 1755 some New Jersey Quakers had their property distrained for refusal to serve in the militia.

It was the Meeting of Ministers and Elders which first sounded the alarm at its General Spring Meeting of March 1755. At that time it addressed an epistle "to Friends on the Continent of America with respect to the commotions and stirrings of the powers of the Earth . . ." The epistle called upon all Friends "to repose . . . in the munition of that rock that all these shakings shall not move" and to maintain a conduct "in all parts of our life as becomes our peaceable profession." It expressed the hope that "nation shall not lift up sword against nation nor learn war any more."[20]

Growing tensions within the Pennsylvania government, together with the increasing anti-Quaker sentiment on the part of the British authorities, prompted the Yearly Meeting the following year to appoint a standing committee to correspond with British Friends and obtain their assistance as the situation might require.

---

20. "An Epistle from our General Spring Meeting of Ministers and Elders for Pennsylvania and New Jersey. . . 1755." Broadside in the Quaker Collection, Haverford College.

Friends in the Delaware Valley

In November 1755 the Pennsylvania Assembly enacted two laws, one to provide £60,000 for "the King's use" to be raised by a special tax, and the other to authorize a voluntary militia. In both cases only a minority of the Quaker assemblymen opposed the legislation. While the bills were under consideration, twenty prominent Friends including Israel Pemberton, John Churchman, John Woolman, John Reynell, and Anthony Benezet submitted a protest against the proposed legislation, but to no avail. This event indicated the widening gap between the nominal Quakers in the Assembly and the spiritual leaders of the Society who were working for a consistent adherence to the ancient testimonies, and at this particular juncture the peace testimony.

Shortly after the passage of these laws, the two new committees of the Yearly Meeting[21] met jointly to consider possible action. Despite long and serious consideration of the new defense measures, no unanimous decision could be reached, and some members ceased attending. Finally twenty-one of the Friends signed and issued unofficially "An Epistle of Love and Caution to Friends in Pennsylvania and New Jersey" in which they declared that "since we cannot be concerned in wars and fightings neither ought we to contribute thereto by paying a tax directed by the said act, though suffering be the consequence of our refusal."[22]

Although the epistle was not an official statement of the Yearly Meeting, its provenance caused it to be seriously received, and it excited much controversy. Consequently the Yearly Meeting in 1757 appointed a committee of thirty Friends to consider the question of paying the tax. Its recommendation that the matter should not be discussed further because of the strong difference of opinion within the Society was adopted by the Yearly Meeting with the exhortation that all Friends should show "charity towards one another."

By now imperial requirements were impinging ever more insistently on colonial affairs. In 1756 the French and British empires were at war, and demands for defense became more imperative. It

---

21. The committee to visit monthly and quarterly meetings and the standing committee to correspond with British Friends.
22. John Churchman, *An Account of the Gospel Labours and Christian Experiences of a Faithful Minister of Christ, John Churchman.* (Philadelphia, 1781), pp. 236-239.

The Founding Years

was therefore inevitable that the attitude of a consistent Quaker assemblyman toward requests for military subventions would create even more tension both within the colony and between the Pennsylvania government and the British authorities. At the same time conditions on the frontier were such that Friends and others who had settled in the area were bound to suffer from the inevitable conflict. Therefore the Yearly Meeting at its sessions that year, on the recommendation of the standing committee for relations with British Friends, converted that committee into a Meeting for Sufferings. It was charged with corresponding with its counterpart in London, with representing the Society where the reputation and interest of Friends were concerned, and with the responsibility for raising and administering relief to Friends on the frontier and elsewhere as needed. Its only limitation was that it "not meddle with matters of Truth and Discipline not already determined in the Yearly Meeting."

The membership of the new Meeting for Sufferings was composed of four Friends appointed from each quarter, to which were added twelve members selected at large by the Yearly Meeting from in or near Philadelphia "for the convenience of their getting together." The various quarters were also instructed to raise a quota of subscriptions for a relief fund and also to keep an account "of any sufferings they may be subject to for the Testimony of Truth," to be submitted to the Meeting for Sufferings as occasion required.

Already Friends in Delaware were subject to a militia law which provided no exemptions, and the Yearly Meeting requested the British Quakers to intercede with the proprietors in support of an earlier appeal it had made directly to them. Subsequently the Meeting for Sufferings approached Lieutenant Governor Denny with an address in behalf of the Delaware Friends.

The attitude of many Friends towards the defense measures taken by the Assembly led to a concerted attack on Quakers both within the province and in England. A law to require a test oath of all members of colonial legislatures was presented to Parliament with the intent of eliminating Quakers from the Pennsylvania Assembly. Some Friends had already withdrawn from serving in the government because of the conflict between service in the legislature and their religious convictions. An important consideration which had

Friends in the Delaware Valley

prevented others from taking a similar step was that thereby the charter guarantees of religious freedom and civil rights might be endangered.

Conditions prevailing in 1756 changed this situation, however. In addition to the outbreak of war with France, the governor, with the support of the Assembly, declared war on the Delaware and Shawnee Indians. Consistent Friends in the Assembly were thus placed in an intolerable position. This, combined with the increasing pressures in England, led both English and American Friends to accept the inevitable. In view of the inconsistency of their role as members of the Assembly, six Quakers headed by James Pemberton resigned their seats, and subsequently several others followed suit, leaving only one-third of the assemblymen Quakers. Quaker control of provincial policy, questionable as it had become, therefore ceased; and although after peace returned seven years later some Quakers returned to sit in the Assembly and exercised considerable influence, they never again held majority control. Those Friends who remained in the Assembly despite strong urgings to withdraw were the subject of persistent and, for the most part, unsuccessful dealings by their respective monthly meetings. Not until the Revolution, however, were Friends disowned for holding positions in the government.

In view of their loss of influence in the government and their deep concern for the situation relative to the Indians, a number of influential Friends headed by the Pembertons decided to form the Indian Association.[23] By this means they were able to work unofficially with the government and with the Indians to bring the conflict with the frontier tribes to an end. The efforts of the Association were instrumental in finally restoring peace on terms acceptable to the Indians.

On the question of personal military service and related activities of any kind, the position of the Society was unequivocal. In 1756 Shrewsbury Quarterly Meeting was informed by the Yearly Meeting that participation in a military watch was military service and that Friends should neither participate nor pay anyone else to do so. Two years later the scope of military service received a further interpretation. In this case some Friends had furnished wagons, horses, and

---

23. Officially, The Friendly Association for gaining and preserving Peace with the Indians by pacific measures.

The Founding Years

drivers for the transport of military goods and gear to frontier posts. The Friends involved gave as excuse the fact that fellow Quakers in the administrative section of the government had contended that such action was not military service. The latter justified themselves by the excuse that it was a requirement of their position in the government.

This situation resulted in a sweeping reinterpretation of the peace testimony as it applied to Friends in government posts as well as to those participating even indirectly in military activities. The Yearly Meeting declared that not only was any assistance to such activities inconsistent with the peace testimony but also that Friends should not continue in or accept any office in government which would oblige them to enforce on their brethren or anyone else compliance with an act which they themselves would scruple to perform. Any Friend persisting in such conduct was to be barred from meetings for discipline or from being "employed in affairs of Truth." This position was strengthened in 1762 by a warning to Friends against "promoting or choosing their brethren to such offices as will subject them to the temptation of deviating from our Christian testimony."

The issue of personally bearing arms arose in acute form in 1764 in Philadelphia Quarterly Meeting which reported to the Yearly Meeting that divers members were being dealt with for such an offense. A group of frontiersmen known as the Paxton Boys, enraged by Indian attacks on frontier settlers, accused the government of failing to furnish defense against the Indians and marched on Philadelphia with the announced intention of massacring Indian refugees sheltered there. The alarm was such that two hundred Quakers took up arms in defense of the city although an outright military confrontation did not occur.

On this occasion the Yearly Meeting spoke in no uncertain terms. It declared that the taking up of arms had brought "calumniation on the Society by their adversaries," and it appointed a committee to consider what action should be taken to make known its concern "to maintain the testimonies of Truth and its Purity and to preserve unity among Friends." Most of the offenders acknowledged their error, although a few persisted in justifying their action as legitimate defense. By the middle of 1767 the matter was finally dropped.

The revitalization of the Society's witness to the testimonies heightened the sensitivity of many Friends to the inconsistency with their Christian principles of holding human beings in bondage. In 1758 Philadelphia Quarterly Meeting called upon the Yearly Meeting to use "every method in our power . . . to put a stop to the practice . . . of buying, selling and keeping slaves." At the same time the epistle from London Yearly Meeting contained a strong indictment of the African slave trade.

Spurred on by a moving appeal from John Woolman, the Yearly Meeting overcame the opposition of the conservative slaveholding element and acceded to the demand for more effective action. It declared that holding slaves was contrary to Friends' principles and directed the monthly meetings to deal with those who persisted in importing, selling, or purchasing slaves and to bar them from sitting in meetings for discipline. It appointed a committee which included John Woolman, John Churchman, John Scarborough, and Daniel Stanton which, assisted by the elders and other faithful Friends in each quarter, was directed to visit all Quaker slaveholders throughout the Yearly Meeting in the cause of the complete abolition of slavery.[24]

As a support to these efforts, in 1759 the Yearly Meeting published Anthony Benezet's *Observations on the Enslaving, Importing and Purchasing Negroes*. A year later it called upon all meetings to discourage slaveholding in hope that in time "the Society will fulfill its goal in this respect." The revised Discipline of 1762 devoted ample space to this concern, and in 1767 the Yearly Meeting directed the quarterly meetings to report the success of "their care therein."

It was not until the next decade, however, that the final step was taken to ban utterly all aspects of the slavery system. In 1774, on the threshold of the Revolution, both Philadelphia and Bucks Quarterly Meetings requested a clarification of the position of the Society on slave holding "for the advancement of our Testimony against a practice so evidently contrary to our Christian Principles and Profession and the Common Rights of Mankind."

In response the Yearly Meeting declared that any Friends concerned with "importing, selling or purchasing or [who] shall give

---

24. When Benjamin Lay, shortly before his death, heard of the decision of the Yearly Meeting he exclaimed: "I can now die in peace."

away or transfer any negroes or other slaves in such manner as their bondage be continued beyond the time limited by law or custom for white persons" should be dealt with, and if they persisted therein they should be disowned.[25] It further directed that such cases should be reported to the next Yearly Meeting, and in the meantime offending members were "not to be employed in the service of truth."

In 1776 the direction for disownment was repeated, and a new query on slaveholding was adopted:

> "Are Friends free of importing, purchasing and disposing of or holding mankind as slaves; . . . and do they use well those who are set free but are not in a position to care for themselves and are they careful to educate them in a religious and virtuous life?[26]

Thus, in the same year as the Declaration of Independence was proclaimed, a struggle for freedom in the Society of Friends had been won. Subsequently the chief concern of the Yearly Meeting was "for the instruction and improvement . . . of those who have been set at liberty . . . (and for) their assistance in worldly concerns."

In the political field the Society was inevitably affected by the burgeoning problems of empire. No sooner was the shock of the Paxton riot affair over when the controversy over Parliament's right to levy colonial taxes erupted, a controversy which was destined to plunge the whole of American Quakerdom into the severest trials of its existence. The Stamp Tax of 1765 and subsequent attempts to derive tax monies from the colonists prompted the colonial merchants to spearhead protest actions which naturally involved most of the Philadelphia Quaker merchants. In early protest activities prominent members of the Yearly Meeting and the Meeting for Sufferings such as John Reynell, treasurer of the Yearly Meeting, James Pemberton, Henry Drinker, and Jeremiah Warder, all members of the Meeting for Sufferings, participated.

An important motive for the involvement of the Quaker merchants in resistance to the policies of the British government, in addition to their strong conviction as to the unconstitutionality of those laws, was their hope to induce moderation and so prevent the protest movement from falling into the hands of the more extreme elements.

---

25. PYM, Min., v.2, pp. 314–315.
26. PYM, Min., v.2, p.355.

Friends in the Delaware Valley

As the leaders of the protests began to organize extra-legal citizens meetings and to bypass government channels in taking action, many Quakers began to fear that adverse changes in the political control of the province might result.

In 1769 the threat of violence on the part of the radical element when the ship *Charming Polly* arrived with a cargo from Britain in contravention of the colonial non-importation agreement, alarmed leading Friends. From this time on the official Quaker policy, determined largely by the more conservative Quakers led by Israel Pemberton, attempted to deter Friends from actively supporting the protest movement. The more liberally minded Friends, such as John Reynell, continued to cooperate with the protesters as long as they conscientiously could.

The attitude of the Yearly Meeting was largely determined by the recommendations of the Meeting for Sufferings which by 1770 had come to view the anti-British current with foreboding. At that time the Yearly Meeting issued a statement advising Friends "to live near The Divine Principle" which will "teach us to avoid mixing with those who are not convinced of our religious principles in their human policy and contrivance." All members were urged to "seek after Quietude and stillness of Mind" in order to be able to advise their brethren who should inadvertently join in and support "by countenance or otherwise the prevailing commotions."

For a few years after Parliament rescinded the Townshend Acts in 1770, there was relative calm. But in 1773 the passage of the notorious Tea Act rekindled the fires of resistance, and there were ominous popular protests in Philadelphia. Realizing the explosive nature of the situation, the Yearly Meeting urged elders and overseers to keep a watchful eye for the maintenance of the peace testimony. It authorized the Meeting for Sufferings to publish such advices as it deemed necessary to guard Friends against "going with the countenancing such measures proposed by the people for asserting their civil rights and privileges" in ways which would not be consistent with the Quaker peace testimony. Subsequently a general meeting of all men Friends in Philadelphia issued advices to the Quaker community "exciting to a due consideration of the nature of our religious profession which requires us to keep quiet and still, both in respect to conversation and conduct." From this point on most Friends avoided participation in the agitations under way.

The Founding Years

The First Continental Congress met in Philadelphia in the fall of 1774 and adopted non-importation, non-exportation and non-consumption agreements against British trade and set up an Association to enforce them. Philadelphia Yearly Meeting which met at the same time issued an epistle to Friends throughout America reminding them that their ancestors had always remonstrated against unjust laws while at the same time avoiding plots and conspiracies. Since Friends in America were indebted to the King for the enjoyment of their liberties, it was their duty to discourage any attempt "to excite disrespect or disaffection to him."

The punitive measures against Boston undertaken by the British government early in 1775 led to an open breach with the colonies, and military action ensued. Some Friends were soon involved, and when the Yearly Meeting met, it received reports of numerous military deviations. Thereupon it issued an urgent warning to Friends in general to avoid all acts, even of being spectators at training grounds, which might compromise their pacifist principles. It also directed the monthly meetings to attempt to retrieve those members who had deviated from their testimony against war and to disown those who persisted in such action.

In October 1775 the Pennsylvania Assembly, despite an appeal from the Meeting for Sufferings, enacted a military enlistment law which provided for the exemption of conscientious objectors on payment of a stipulated sum. About the same time the Continental Congress urged the abolition of the old colonial governments, and the Quakers became alarmed as to the fate of their provincial charter. They were also opposed to independence if it were brought about by war. Therefore in January 1776 the Meeting for Sufferings issued "The Ancient Testimony and Principles of the People Called Quakers renewed with respect to the King and Government and touching the Commotions now prevailing in these parts of America." In addition to stating the pacifist position of the Society, the "Ancient Testimony" admonished Friends that it was not their business to join in plots and contrivances for the setting up and pulling down of kings and governments which alone was God's prerogative.[27]

Within a few months independence was declared, and the Yearly

---

27. PYM Meeting for Sufferings, Min., v.2, pp. 53-59.

Friends in the Delaware Valley

Meeting of 1776 faced a new and urgently serious situation, especially since the position taken by the Quakers had created great animosity in many quarters. Representatives were present from all the other American Yearly Meetings except New York where the military conflict then centered. The recommendations of an inter-yearly meeting committee as to the conduct required of all Friends under the prevailing circumstances were adopted by the Yearly Meeting. This statement enjoined Friends from participating in the new civil governments inasmuch as they were being supported by war. Friends were forbidden to pay any fine, penalty, or tax in lieu of military service for themselves, their family members, or their servants; they were not to engage in any business likely to promote the war; nor were they to buy or sell prize goods as spoils of war. As a positive expression of their principles they were urged to provide for the distress and suffering of those adversely affected by the war.

Throughout the conflict, wartime legislation which bore heavily on the Quakers included tax laws and loyalty test laws, the latter at one time barring non-jurors from most professions and from maintaining or teaching in schools. Despite the heavy penalties for refusal to comply with loyalty test laws the Yearly Meeting firmly declared that Friends could not comply with such requirements nor pay fines for such non-compliance. The payment of war taxes was equally banned, although the payment of mixed taxes, to be used for both civil and defense purposes, was left to the individual conscience, as was the acceptance of the new continental currency.

A particularly disturbing event for the Friends was the appearance in August 1777 of a spurious document purportedly issued by the Spanktown Yearly Meeting detailing military information for the British. Although the document was clearly fraudulent, the immediate effect of its disclosure was widespread apprehension among the patriots as to the true intentions of the Quakers. Consequently a number of leading Friends and other citizens were arrested. Within a short time nineteen Friends including James, John, and Israel Pemberton, Henry Drinker, Owen Jones, Jr., and John Hunt were exiled to Winchester, Virginia, where they passed the winter of 1777–1778.

When soon afterward the Yearly Meeting assembled, it issued a strong defense of the Society, branding the Spanktown document as utterly false and declaring that the apprehended Quakers were innocent of any offense and should be freed. A deputation from the Yearly Meeting presented this statement to both Generals Washington and Howe. It was not until the next spring, however, that all but one of the exiled Friends were liberated, John Hunt having died while in detention.

Throughout the war the Meeting for Sufferings consistently appealed to the authorities on behalf of Friends imprisoned for their stand against any participation in the war effort. It also defended the Society against many unfair and unfounded charges levelled against it. In September 1779 the meeting issued a statement against such an attack declaring that "great industry had been used by Persons of a malevolent Disposition to kindle jealousies, by insinuations against our religious Society."

Many Friends, especially younger ones, were swept into the war, strongly attracted by its characterization as a struggle for freedom. Peripheral aspects related to the war such as payment of taxes or monetary penalties for refusal to conform to wartime legislation brought disciplinary action on many older Friends sympathetic to the colonial cause. In the course of the war 1,276 members were disowned, 758 for military deviations, 239 for paying fines and taxes, 125 for subscribing loyalty tests, 69 for assisting the war effort, 32 for accepting public office, and 42 for miscellaneous deviations such as watching training exercises, celebrating independence, etc.[28]

Although the military phase of the war ended in October 1781, wartime problems for the Friends did not cease immediately, and prosecutions for refusal to pay war taxes brought heavy trials for many even after the peace of 1783. Distraints of property resulting from opposition to the war amounted to £56,767 for members of the Yearly Meeting, £38,550 for Pennsylvania, £16,027 for New Jersey, and £2,190 for Friends in Delaware. The system of distraints was especially onerous since forced sales often resulted in low returns, and frequently prejudice against the Quakers was expressed through this procedure. In the eyes of those working for the purification of the life of the Society, however, the various types of suffering,

---

28. Arthur J. Mekeel, *The Relation of the Quakers to the American Revolution*. (Washington, D.C., 1979) pp. 200–201.

imprisonment, exile, personal abuse, and extensive property losses were welcomed as a means of loosening the Friends from attachment to worldly possessions.

Among those who were disowned for their refusal to accept the strict interpretation of the peace testimony as maintained by the Yearly Meeting were such persons as Isaac Gray who published an unauthorized pamphlet justifying the payment of all taxes, Clement Biddle who became quartermaster under General Gates at Valley Forge, and Samuel Wetherill, Jr., a minister. These and other like-minded former Friends banded together in 1781 to establish an independent group of Friends called the Free Quakers. The Free Quaker Discipline dispensed with disownment and was far less precise in regard to particular testimonies. A meeting house, still standing, was built in 1783 at the corner of Fifth and Arch Streets, and the meeting continued to function for a half century with dwindling numbers until it closed its doors in 1836.

An important aspect of the Quakers' role in the Revolution was their work in supplying relief for those suffering the effects of the war. Especially notable was the relief project in cooperation with New England Friends in behalf of the besieged city of Boston and the surrounding towns in the winter of 1775-1776. Ultimately over £3,000 were raised by the Yearly Meeting for this purpose. As the war progressed, such activities extended to Virginia and the Carolinas, as well as including their compatriots at home. In 1790, on the floor of the first federal Congress of the United States, Elias Boudinot paid eloquent tribute to the relief activities of the Quakers when he said:

> Many Quakers of this city (Philadelphia) exercised such humanity . . . as did honour to human virtue. The miserable prisoner not only felt the happy effects of their exertions in his favor, but participated in their money, their food, and clothing.[29]

The impact of the Revolution in provoking extensive deviations from the Quaker peace testimony spurred the Friends in their efforts to revitalize the spiritual life of the Society, especially as evidenced in a strict adherence to the testimonies and the distinctive Quaker way of life. In 1777 the Yearly Meeting appointed a committee to visit the

---

29. Elias Boudinot, *The Life, Public Services, Addresses and Letters of Elias Boudinot,* ed. J.J. Boudinot, 2 vols. (Boston & New York, 1896), v. 2, p.227.

The Founding Years

quarterly and monthly meetings to assist in visiting families "for a Reformation loudly called for." In the belief that "much corruption hath been introduced amongst us . . . in . . . mixing in Schools for our Children where there has not been a regulation consistent with our Holy Profession and the Principles of Truth," it warned against sending the children of Friends to such schools and urged that schooling be provided under Quaker auspices.

The following year the Yearly Meeting urged the subordinate meetings to establish schools "for the instruction of our children in useful learning" and it appointed a committee to assist the quarterly and monthly meetings in this endeavor. It also recommended that the quarterly and particular meetings collect a fund for this purpose and that a query be instituted on this subject to keep the Yearly Meeting aware of developments and problems.

The success of these efforts was evident in reports the next year that several monthly meetings had set up schools and others were in the process of doing so. In 1783 the Yearly Meeting again brought to the attention of Friends the minute of 1778 on the education of their children and urged them to continued action. Four years later in 1787 all but one of the quarterly meetings reported the promotion of schools pursuant to the advices of the Yearly Meeting.

Another area of special concern was the distillation and consumption of spirituous liquors which, although not new, was becoming more insistent in the efforts to avoid practices harmful to the achievement of a non-worldly Quaker way of life. Already in 1774 Anthony Benezet had portrayed the evils of intemperance in *The Mighty Destroyer Displayed* which he followed in 1778 with *Remarks on the Nature and Bad Effect of Spirituous Liquors,* tracts which doubtless had an influence on the attitude of many Friends. In 1777 Western Quarterly Meeting expressed its conviction that Friends should not be involved in the production of liquor "and in particular the destroying of grain for that purpose." The Yearly Meeting consequently called upon the monthly meetings to caution their members against "distilling, buying or selling grain for that purpose" or using or partaking of liquor made of grain. Coupled with this exhortation was an admonition against keeping taverns and public houses of entertainment which dispensed such spirits.

In 1782 the Yearly Meeting revised query four to read: "Are

Friends careful to discourage the unnecessary distillation or use of spirituous liquors?"[30] In 1784 and again in 1788 the yearly meeting urged against the importation and retailing of spirituous liquors. In the latter year it recommended that all quarterly and monthly meetings appoint a committee "to visit and treat with members" who imported distilled liquor from the West Indies or elsewhere as well as with those who distilled liquor from grain or other produce. Not until the next century, however, did the Friends come to the point of advocating total abstinence.

Although refraining from participation in politics and government, Friends took an active part in the moral and social issues of the day. With the end of the Revolution in 1783 the Society moved on from freeing its own slaves to a frontal attack on the institution in general. Spurred by the action of the British Quakers in appealing to Parliament for the suppression of the African slave trade, the Yearly Meeting submitted an "Address of the People Called Quakers" to the United States Congress, calling for the prohibition of the further importation of Negroes from Africa as slaves. Persistent efforts in this cause were to be one of the chief activities of the Friends as citizens of the new nation. The following year the Meeting for Sufferings presented a copy of the address of the British Friends to Parliament to both the General Assembly of Pennsylvania and the Council and Assembly of New Jersey.

Two years later, in 1786, the Meeting for Sufferings sent a committee which included Henry Drinker and James Pemberton to New York, then the seat of the national government, with a similar address from the meeting itself to the United States Congress. Although Congress was not in session, the committee interviewed John Jay, Minister of Foreign Affairs, and various members of Congress, finally leaving the address with the New York Meeting for Sufferings for future action.

Continuing its active concern on this issue, in 1788 the Meeting for Sufferings appealed successfully to the legislatures of Pennsylvania, Delaware, and New Jersey for laws prohibiting the current practice of fitting out ships for the slave trade in the ports of those states and the forcible transportation of freedmen to the West Indies for sale

---
30. PYM, Min., v.3, p.50.

The Founding Years

into slavery. A year later the Yearly Meeting presented an address to the new United States Congress urgently calling for action by the federal government to stop the slave trade.

By the time of the establishment of the new national government in 1789, an extended era of peace had already opened and the Friends had found an accepted place in their respective communities. The Yearly Meeting took the occasion to present an address to President Washington assuring him of their firm loyalty and pledging to contribute freely to the necessary support of civil government and to the exigencies of the poor. Washington replied with a gracious statement of his high regard for the Quakers and the constructive role which they played in society.

Friends in the Delaware Valley

Bibliographic Suggestions

General Works:

Standard older works are Isaac Sharpless, *A History of Quaker Government in Pennsylvania,* 2 vol., Philadelphia, 1902, and *Political Leaders of Provincial Pennsylvania,* New York, 1919. The fruit of more recent scholarship which includes social and economic aspects is Frederick B. Tolles, *Meeting House and Counting House, The Quaker Merchants of Philadelphia, 1682-1763,* Chapel Hill, 1948.

Special Topics:

The Founding Years:

Bronner, Edwin B., *William Penn's "Holy Experiment,"* New York, 1962.

Dunn, Mary Maples, *William Penn, Politics and Conscience,* Princeton, 1967.

Lokken, Roy, *David Lloyd, Colonial Lawmaker,* Seattle, 1959.

Nash, Gary B., *Quakers and Politics, Pennsylvania, 1681-1726,* Princeton, 1968.

The Keithian Controversy:

A standard biography of George Keith is Ethyn Williams, *George Keith, 1638-1716,* New York, 1942. A good brief account of Keith's life is James Arthur Muller, "George Keith" in the *Historical Magazine of the Protestant Episcopal Church,* v.13, pp. 94-106.

Recent scholarly treatments of Keith are J. William Frost, *The Keithian Controversy in Early Pennsylvania,* Norwood, Pa., 1980, and his excellent discussion of the Keithian controversy, "Unlikely Controversialists: Caleb Pusey and George Keith," *Quaker History,* v.64, pp. 16-36.

The Middle Years:

Brock, Peter, *Pioneers of the Peaceable Kingdom,* Princeton, 1968, Chap. 2, 3.

Rothermund, Dietmar, *The Layman's Progress, Religious and Political Experience in Colonial Pennsylvania, 1740-1770,* Philadelphia, 1961.

Thayer, Theodore, *Israel Pemberton, King of the Quakers,* Philadelphia, 1943.

The Founding Years

_____, *Pennsylvania Politics and the Growth of Democracy, 1740-1776,* Harrisburg, 1953.

Tolles, Frederick B., *James Logan and the Culture of Provincial America,* Boston, 1957.

_____, *Quakers in the Atlantic Culture,* New York, 1960.

The Reform Movement:

Tolles, *Meeting House and Counting House,* chap. 10.
James, Sydney V., *A People Among Peoples,* Cambridge, 1963, chap. 9.

The Revolutionary Period:

An older standard work is Isaac Sharpless, *History of Quaker Government,* v.2. More recent accounts are Arthur J. Mekeel, *The Relation of the Quakers to the American Revolution,* Washington, D.C., 1979, and Richard Bauman, *For the Reputation of Truth,* Baltimore, 1971. Also see Brookes, George S., *Friend Anthony Benezet,* Philadelphia, 1937.

Friends and Slavery:

Good surveys of the subject are in Thomas E. Drake, *Quakers and Slavery in America,* New Haven, 1950, chap. 1-5, and Sydney V. James, *A People Among Peoples,* chap. 7, 8, 12.

J. William Frost, "The Origins of the Quaker Crusade against Slavery, A Review of Recent Literature," *Quaker History,* v. 67, pp. 42-58, gives an excellent analysis of the various viewpoints on the subject.

Development of the Discipline:

A brief analysis of the changes in the Book of Discipline during the period is Raynor W. Kelsey, "Early Books of Discipline of Philadelphia Yearly Meeting," *Bulletin of the Friends Historical Association,* v.24, pp. 12-23. A much more extensive study is J.P. Marietta, *Ecclesiastical Discipline in the Society of Friends, 1682-1776,* Ann Arbor, 1971.

Education:

Early works are Thomas Woody, *Early Quaker Education in Pennsylvania,* New York, 1920, and *Quaker Education in the Colony and State of New Jersey,* Philadelphia, 1923. For a recent discussion of the subject see J. William Frost, *The Quaker Family in Colonial America,* New York, 1973, chap. 5, 6.

MEETING-HOUSE, FOURTH & ARCH STS.

Arch Street Meetinghouse, Fourth and Arch Streets, built by the yearly meeting (1804–1811) and used by it until 1827. During the Separation the Orthodox used this building and since 1955 it has been the site of the united yearly meeting.

II Years of Crisis and Separation: Philadelphia Yearly Meeting, 1790–1860

by J. WILLIAM FROST

The Yearly Meeting at the Beginning of the Nineteenth Century

In 1808 Philadelphia Yearly Meeting witnessed the retirement of an individual who had symbolized the Society since the 1750's. James Pemberton (1723–1809), was a founder of the Pennsylvania Hospital, member of the Pennsylvania Assembly who gathered grain for Braddock's expedition and resigned in the Quaker withdrawal from government, member of the Meeting for Sufferings since 1756 and often clerk of the organization, clerk of Philadelphia Yearly Meeting, exiled during the Revolution, President of the Pennsylvania Abolition Society from 1790, member of the committee to issue the first printed discipline in 1797, and carrier of frequent Quaker petitions to the Pennsylvania legislature, the colonial and revolutionary governments, and finally the Congress of the United States. Also stepping down was Henry Drinker (1734–1808), slightly younger than Pemberton, who first became famous as owner of a ship carrying taxed tea in 1772. Drinker was a prominent merchant, on the Board of Overseers of the Friends' Public School of Philadelphia, member and later clerk of the Meeting for Sufferings, exiled to Virginia during the Revolution, member of the American Philosophical Society, elder of Philadelphia Monthly Meeting, and treasurer of the Yearly Meeting.

Friends in the Delaware Valley

If the two men were physically able to attend the 1808 sessions, they could have commented upon the extraordinary continuity in forms of worship and organization. As in the 1740's the Meeting of Ministers and Elders came the preceding day; the Yearly Meeting still opened with the accounting of representatives, reading of epistles, and reports of the quarterly meetings. The delegates from quarterly meetings caucused and selected a clerk and his assistant, who were then confirmed by the Yearly Meeting. Generally the same men served many years. Committees dealt with issues raised by the quarterly meetings and adjudicated appeals for disciplinary cases decided by monthly meetings. The proceedings of the Meeting for Sufferings were read and approved, epistles written, and policy decreed. Some problems seemed perennial: midweek meetings were poorly attended, drowsiness in some meetings was too prevalent, and children occasionally deviated from plainness. Talebearing was discouraged, and love and unity generally prevailed except in a few instances.

Between 1758 and 1808 the central concerns of the meeting had not changed. First came worship, as men and women sitting in separate meetings waited for the direct inspiration of the Lord before speaking. The Yearly Meeting occasioned an immense gathering of Friends from Pennsylvania, New Jersey, and Delaware, with visitors from England and throughout America. Quakers communicated in a special language, using distinctive terms like "being owned by the Lord," "weighty," and "centering down," which set them apart from other religious communities and insured communication between Friends old and young, near and far. The social testimonies of 1758 were still current, though with some changes due to external environment: slavery and the incorporating of blacks into American society, justice to and civilization for the Indians, moderation in outward comforts, resistance to any form of military service, education for Quaker children, charity to the poor, justice and fairness in business enterprises.

Pemberton and Drinker had been involved in the end of the Holy Experiment during the Revolutionary crisis. Before 1776 Quakers shaped the destiny of Pennsylvania in the assembly and, after 1757, especially through lobbying with Parliament, either directly or in concert with English Friends. Now Friends were a minority denomination, though still with many socially prominent and wealthy members, who petitioned the state legislature in Harrisburg, and the Congress, no longer in Philadelphia but now in Washington, D.C. The discipline advised Friends to hold no appointive or elective political office, frowned on membership in political parties, and approved of the reluctance of many even to vote. When in the 1790's speculation in federal government securities was rife due to Alexander Hamilton's plan to fund debts growing out of the Revolutionary War, the Yearly Meeting ordered Friends to avoid any profits coming from war and to refrain from buying such bonds.[1] By the 1820's a few Quakers gradually made their way back into political affairs, but the meeting continued to look askance at such worldly activities.

The loss of governmental power did not diminish the attractiveness of the Society of Friends, and Philadelphia Yearly Meeting probably attained its largest membership in the period before 1820. Rarely was there a year when reports of quarterly meetings did not list new meeting houses, additional meetings for worship, and even reorganized monthly meetings. Because of the press of business, Philadelphia Monthly Meeting divided into Northern and Southern Districts in 1772; later Philadelphia Quarterly Meeting was split with the more rural sections being organized as Abington Quarterly Meeting. Women Friends found their quarters at yearly meeting sessions cramped, and so, after years of discussion, a new meeting house designed for special activities of Philadelphia Monthly Meeting and the Yearly Meeting was erected in 1804 over the Quaker graveyard at Fourth and Mulberry. Arch Street Meeting House, as it is now called, like many eighteenth and early nineteenth century meeting houses in Crosswicks, Darby, Falls, Gwynedd, London Grove, Providence and elsewhere, provides a striking testimonial to the Quaker esthetic of this period.

In 1790 meetings on the eastern shore of Maryland, including Third Haven, Cecil, and Duck Creek were transferred to Philadelphia Yearly Meeting and new meetings west of the Susquehanna (Menallen, Warrington, Centre, etc.) became part of Baltimore Yearly Meeting. The great westward migration which decimated meetings in Virginia, North Carolina, and Nantucket had less impact

---

1. Philadelphia Yearly Meeting, Minutes, 1791, pp. 202, 211.

Friends in the Delaware Valley

on Philadelphia Friends. Occasionally one finds a certificate of removal for a New Jersey Quaker in eastern Ohio, but Philadelphia Yearly Meeting remained secondary to Baltimore in sponsoring the new monthly meetings at Redstone, Barnesville, and Mount Pleasant.[2] Philadelphia was consulted, however, when Ohio Yearly Meeting was formed in 1812 with a membership of nearly 20,000.

Between 1800 and 1860 life in the Philadelphia region became transformed by the beginnings of American industrialism. Chester, Bucks, and Burlington counties more than doubled their populations in sixty years while Philadelphia included 54,000 inhabitants in 1790, 188,000 in 1830, and 565,000 by 1860. Commerce in colonial days consisted mostly of the export of agricultural or semi-finished goods, and artisan manufactures destined for American markets. In 1783 there was only the Bank of North America; by 1860 the first and second Banks of the United States had come and gone, and Chestnut Street was famous for banks and insurance companies, the stock exchange and the corn exchange. Turnpikes and canals were being superseded by railroads; steamboats were making sailing ships uneconomical.

How did Friends respond to this economic transformation? Many prospered as land values in the city rose and farmers close by sold to the enlarged urban market. While some Friends, and even on one occasion the Yearly Meeting,[3] denounced paper wealth, many Quakers participated in the new financial and marketing institutions and invested in canals and railways. Josiah White (1781–1850), through a new kind of lock, made the Lehigh River navigable, pioneered the use of anthracite coal for home heating, and fostered the development of the Pennsylvania iron industry by using anthracite for smelting. Joseph Wharton (1828–1906) laid the foundation for his own fortune and the American lead and zinc industries before 1860. Jacob Ridgway (1768–1843), who left an estate of $3.5 million, began in the wholesale grocery trade and expanded into importing and real estate. Mordecai Lewis's ships traded with India and China as well as Europe; Samuel Logan Fisher added to his fortune through iron works; J. B. Lippincott was a publisher; Joseph Lover-

---

2. James Harris Norton, Quakers West of the Alleghenies and in Ohio to 1861, Ph.D. diss., Western Reserve University, 1965, pp. 32–34, 54.
3. PYM, Minutes, 1805.

ing was in sugar refining; John Sellers in flour mills; Thomas Gilpin established a paper factory and cotton and woolen mills on the Brandywine. Quakers established important retail stores for groceries, dry goods, clothing, and drugs.[4] In spite of their varied financial activities, the increased size of Philadelphia and its manufacturing establishment meant that Quaker businessmen became less important to the region than in colonial days.

Eighteenth century Quaker merchants demonstrated civic consciousness by supporting a variety of institutions: the Pennsylvania Hospital, the Library Company, the Alms House, and fire companies. Nineteenth century Quakers continued this tradition of benevolence. Roberts Vaux (1786-1835) helped establish the Philadelphia public school system and special schools for the blind and the deaf and dumb, the Athenaeum, the Apprentices Library, the Historical Society of Pennsylvania, the Philadelphia Vaccine Society, and the Academy of Natural Sciences. Vaux worked in programs for the improvement of farming methods, in creating the Pennsylvania Prison Society and visiting prisoners, in the Pennsylvania Abolition Society, and in temperance. Vaux, who had inherited wealth, made reform his vocation and even entered into politics as a strong Jacksonian and opponent of banks.[5] More typical was Thomas Pym Cope (1768-1854) who combined business acumen with welfare and meeting activities. Cope, whose packet lines competed with the ships of Stephen Girard, served as a manager of the Mercantile Library, on the Board of Trade, and as a trustee of the Pennsylvania Hospital, the Friends Asylum in Frankford, Haverford School, and the Friends Bible and Tract Societies.[6] A very different, and much more radical reformer was Lucretia Mott (1773-1883). Clerk of the Women's Yearly Meeting, traveling minister, member of the Philadelphia Female and American Antislavery Societies, supporter of the Women's Medical College of Philadelphia, a founder of Swarthmore College, and the most famous feminist of her generation, she served

---

4. Eleanor Morton, *Josiah White* (New York, 1946); Ross Yates, "Samuel Wetherill, Joseph Wharton and the Foundation of the American Zinc Industry," *Pennsylvania Magazine of History and Biography*, 98 (Oct., 1974), pp. 469-514.

5. R. N. Ryon, Roberts Vaux: A Biography of a Reformer, Ph.D. diss., Penn State University, 1966.

6. Eliza Cope Harrison, ed., *Philadelphia Merchant: Diary of Thomas Pym Cope* (South Bend, Indiana, 1978), p. 19.

as a publicist for a wide range of reforms and associated with the most important men and women of America.[7]

Philadelphia Yearly Meeting showed reluctance to sponsor charities aimed primarily at outsiders who were not black or Indian. After receiving a bequest for the education of the poor of all religions in a school to be controlled by trustees appointed by Friends, the Yearly Meeting declined involvement.[8] It also refused to govern the Friends Asylum and suggested instead a self-perpetuating group of Quaker trustees.[9] The Asylum at Frankford, patterned after the Retreat in Yorkshire founded by Samuel Tuke in 1796, opened in 1817. Following Tuke's recommendations, the Asylum emphasized a serene, beautiful environment and moral responsibility rather than close confinement and strong medicines. The Asylum treated only Quakers, partially so that a "family" atmosphere stressing responsibility, acceptance, and manual labor could emerge. The number of patients remained small until after the separation when, under Orthodox control, the institution began to admit outsiders. As a private hospital with relatively expensive fees, Friends Asylum catered largely to middle and upper class patients and attempts to provide subsidies for poor Friends failed. Treatment in the hospital before the Civil War combined, in general, the prevailing canons of American medicine with Tuke's procedures. In 1851 the hospital claimed a recovery rate of 72 percent.[10]

At the close of the Revolution the Meeting for Sufferings assumed the responsibility for the supervision and distribution of publications. Reprints of works of early Friends and of contemporary English Quakers, epistles from London and Philadelphia Yearly Meeting, and special tracts on slavery or temperance were given or sold to each meeting and to Quakers in remote areas. The size of each edition was normally from 1,000 to 3,000 copies. Tract societies in England and America, founded at the beginning of the nineteenth

---

7. Margaret H. Bacon, *Valiant Friend: The Life of Lucretia Mott* (New York, 1980).
8. PYM, Minutes, 1816, p. 274.
9. *Ibid.*, 1812, pp. 212-213.
10. Kim Van Atta, *An Account of the Events Surrounding the Origin of Friends Hospital* (Philadelphia, 1976): Norman Dain and Eric T. Carlson, "Milieu Therapy in the Nineteenth Century: Patient Care at the Friends Asylum, Frankford, Pennsylvania, 1817-1861," *Journal of Nervous and Mental Disease,* 131, No. 4 (Oct., 1960), pp. 277-290.

century, published large quantities of cheap materials; English Quakers had three such societies before the Friends Tract Association was formed in 1817. This association, like the Asylum, was an independent organization formed and controlled by members of Philadelphia Yearly Meeting. During its first year the society printed 47,028 copies of such tracts as *Memoirs of John Woolman* and *Little Sins.* The Tract Society, which continued under the control of the Orthodox, claimed to have issued 332,000 pamphlets by 1830. For 1841 alone, the annual report listed 144,183 copies printed and 120,274 distributed to the West Indies, the Western Soup House, the penitentiaries, and the Magdalen Asylum.[11] Through its distribution of almanacs, calendars, sketches of prominent Friends, and disquisitions on doctrine, the Tract Association served as a major vehicle for propagation of the ideas of the Society of Friends.

At the retirement of James Pemberton and Henry Drinker, the continuity in Philadelphia Yearly Meeting had existed for 130 years and there was no reason to assume that the future would be more traumatic than the past. Yet within ten years tensions within the Society would begin to build into a controversy which would disrupt and change the relation of Friends to each other and to American society. The next sections of this chapter will consider the long term and immediate causes, personalities and events, interpretations and eventual results of the Hicksite-Orthodox schism. By 1840 the Society would be divided into two warring and four suspicious factions. While Quakers might despise other Quakers, their similarities in dealing with problems of education, temperance, and slavery proved how alike they still were.

Separation

The search for continuity in the Society of Friends is a fascinating as well as frustrating task. In no other denomination can one generation's emphases be so completely transformed by the next, while both continue to define themselves as remaining faithful to the

---

11. Edwin B. Bronner, "Distributing the Printed Word: The Tract Association of Friends 1866-1966," *Pennsylvania Magazine of History and Biography,* 91 (July, 1967), pp. 342-354.

essence of the faith and in an unbroken tradition of continuing revelation. None of the theological issues confronting Friends in the early nineteenth century was new and most had been debated at length during the formative years before 1690 when the acceptable perimeters of belief were delineated. Yet disagreement came on fundamentals: what was the Inward Light and how was it connected with Jesus of Nazareth, what was the relationship of intellect and reason to the operation of the Inward Light, what authority did Scripture possess and what power did it convey, how much uniformity in belief was to be required, and who guaranteed the transmission of the authentic faith? Not ethics or ritual but the relation of belief to contrasting experiences of either radical inner obedience or self-surrender to the atonement came increasingly to be seen as decisive, and eventually became divisive.[12]

Quakers have always justified their distinctive doctrines within the prevailing thought patterns in the general culture. In the eighteenth century a dominant feature in Christian thought was rational religion. The essence of Christianity was assent to a short and simple set of propositions about God, Jesus, and the afterlife and the practice of morality. The impact of rational Christianity upon Quakers remains problematic, partially because no precise lines had to be drawn. Friends believed that the Inward Light was the working of Christ or the Holy Spirit within the consciences of all human beings. If non-Quakers tended to deify or sacralize reason and then confuse it with the light, members did not strenuously object. In their desire to emphasize the mercy of God to everyone, Quakers tended to play down but never deny the necessity of historic revelation.

Even when evidence has been marshalled to show the easy compatibility of some elements of Quakerism with natural religion, one must also stress the basic incongruity of many beliefs and practices.

---

12. The primary and secondary literature on the schism is immense. In addition to minutes and correspondence, there are periodicals and tracts, the transcript of a trial in 1832 when leading Friends on both sides testified on the events, and various narratives. William Hodgson, *The Society of Friends in the Nineteenth Century* provides a Wilburite perspective; Samuel Janney, *History of the Religious Society of Friends,* IV gives a Hicksite view; Allen C. Thomas, *A History of the Society of Friends in America* is moderately evangelical. Edward Grubb, *Separations and Their Causes* provides an English Quaker perspective. The best short summary remains Elbert Russell, *History of Quakerism,* chapter 22.

Years of Crisis and Separation

The entire system of Quaker worship and peculiar customs—pacifism, endogamous marriage, discipline by monthly meetings—depended upon quietist presuppositions. Quaker quietism rested upon the belief that any involvement of the human will, reason, emotions, and intellect contaminated the experience of the Inward Light. Quietism had roots in the attack by George Fox and early Friends upon Puritan scholasticism and their reliance upon direct revelation or "openings" as the basis for ministry. In the eighteenth century the emphasis upon God's direct contact with the believer had become so pronounced that all acceptable preaching was deemed to come from a "feeling" of God's presence. Those who accepted the divine promptings and preached as they were prompted were accorded extraordinary deference by the membership, but ministers' journals show the psychic cost in their dread of presuming to speak for God when self-will might become dominant at any instant.

Quaker quietism is best illustrated through the *Journal of John Woolman.* Woolman defined the essence of religion as pure truth or wisdom, and quoted Scripture to illustrate true obedience. The purity of inner revelation was seen as the same whether in Roman Catholic, Anglican, Indian, or Quaker. He made no attempt to justify the truths of religion but assumed an implicit agreement between Quakerism and Christianity and between Christianity and "Pure Wisdom." Reconstruction of Woolman's library shows that he was extraordinarily well read in the apologists of rational religion like Locke and Bishop Tillotson, but he neither endorsed nor rejected their beliefs. There is no passage in Woolman on the significance of the virgin birth, the outward atonement, the trinity, or the Bible. Woolman agonized over slavery, the incessant pursuit of wealth, the reliance of farmers on alcohol, resistance to taxes for war, and the ill effects of war. His solution to all social problems was the same: purify the soul of evil, and harmony with man and God will result.

In the early nineteenth century Quakers on both sides of the Atlantic in significant numbers modified or repudiated quietism.[13] In England and America the French Revolution symbolized the excesses

---

13. Elizabeth Isichei, *Victorian Quakers* (London, 1970) is the standard source on English Friends in this period. The Records and Recollections of James Jenkins, ed. J.W. Frost provides an observer's account on changes in English Quakerism.

to which a worship of reason and attack upon revelation could lead. No longer seen as a perfect guarantor of truth, quietism seemed inadequate to buttress the faith against the challenges of the enlightenment or an increasingly negative rationalistic religion. The effrontery of a minister presuming to speak for God and then misquoting Scripture, prophesying inaccurately, using bad grammar, or showing the effects of indigestion made educated Quakers search for a more factual grounding for their faith.

The decline of quietism and rationalism in relation to rising evangelicalism appeared in both England and America. In 1783 the largest church bodies in America were Presbyterian, Congregationalist, Anglican, and Quaker. By 1850 the Methodists and Baptists had numerical dominance and the Presbyterians had been able to continue their growth only by surrendering to the revival techniques associated with the Second Great Awakening. The rapid increase of immigration, the loss of political power and prestige, and the undynamic quality of traditional Quaker worship stimulated prominent Philadelphia Friends to rethink the implications of their faith. We cannot now determine whether those American Friends (later termed "Orthodox") became evangelical because English Friends had first embraced these doctrines, because of the pervasive effects of evangelical doctrines in American society, or because they were in search of a method of revitalizing Quakerism and the successful theology at hand was evangelicalism. What may seem difficult for us to understand is that evangelicalism was adopted by many of the most educated and cosmopolitan Friends. Methodist and Baptist evangelicals flourished in camp meetings in the South, the frontier, and in the revivals of the Burned Over District of Upstate New York. But, evangelicalism also flourished at Yale College under President Timothy Dwight, and Jedediah Morse led an awakening in Boston designed to counter the pernicious influences of rational theology at Harvard. Quaker evangelicalism is more closely related to the sophisticated expositions of the faith by London Quakers or Connecticut Congregationalists than to the backwoods extravaganzas of Methodists and Baptists.

While the immediate causes of the schism among Friends were theology and church government, the patterns of division were influenced by geography, economic status, and kinship.

Years of Crisis and Separation

Robert Doherty[14] charted a profile of the Orthodox within the city of Philadelphia and concluded that they composed a socially cohesive, business oriented, small community. Those in the country who joined them had close contacts with Orthodox leaders, family connections being very important. Hicksites in the city were likely to be recent migrants from rural areas or families with old wealth like the Biddles and Whartons who resented the commercialism of the rising entrepreneurs. There was a high correlation betwen evangelical emphasis, residence in a city, ownership of stock, upward mobility, and participation in commercial enterprises. Evangelicals tended to favor close cooperation with other religious groups in a great enterprise to make America a Christian civilization. Quaker evangelicals in England and America belonged to Bible societies, created tract societies, endorsed missionary work, and supported a variety of reform groups advocating temperance, prison reform, peace, and antislavery. Evangelical religion among Quakers is notable for an intense activism.

Historians attempting to explain why evangelicalism appealed to conservative merchant classes have developed several intriguing hypotheses. One interpretation stresses the ambivalence of Americans wanting wealth yet desirous of a much simpler life than commercialism represented.[15] Those who were prospering in the rapid expansion of Philadelphia desired, unconsciously perhaps, a religion which would enable them to escape from guilt. Supposedly, quietism with its demand for purity and "centering down" was less attractive than evangelicalism with its emphasis upon correct belief, benevolent activities, and conversion.

An alternative explanation views the Quaker merchants as struggling under the dislocation of rapid economic transformation and searching for a means of stability. Evangelical doctrines which stressed the authority of Scripture, historical revelation, and agreement upon the fundamentals of Christianity offered a much more easily attainable certainty than had traditional quietism. Evangelicalism also offered a method of controlling the excesses of American democracy, for the network of benevolent voluntary associations

---

14. Robert Doherty, *The Hicksite Separation* (New Brunswick N.J., 1967).
15. Marvin Myers, *The Jacksonian Persuasion* (Stanford, 1957).

which would Christianize the workers and the immigrants would also guarantee the future of American democracy by producing the educated moral citizenry necessary for the functioning of society.[16]

One difficulty with accepting these explanations is that they ignore why people said they were acting as they did. We can establish a correlation between evangelical doctrines and new wealth, but we cannot prove that evangelicals were more insecure and more worried about America than liberals or quietists. Sufficient numbers of country dwellers became Orthodox and enough city merchants became Hicksites to make any psychological interpretation suspect. A more plausible hypothesis is that evangelicalism was used by part of the leadership of Philadelphia Yearly Meeting in an attempt to revitalize the Society of Friends. In London an older traditional quietist leadership was gradually replaced, and although there were serious conflicts there was no major separation. Urban Philadelphia Friends who became Orthodox were more likely than were the Hicksites to have visited England and to have entertained English Friends in their homes. Since Philadelphia Quakers, even while claiming equality of yearly meetings, stood very much in awe of English respectability, the fear of heresy and advocacy of evangelical doctrines by traveling English ministers were powerful influences. City Friends may have felt the growing numerical insignificance of Quakers more than country dwellers. Compared with other Protestant groups, Friends were more and more becoming a minority and exercising less impact upon an area which they felt responsible for founding and forming. Quaker ministers and elders were attracted to the evangelical doctrines espoused by New School Presbyterians and Methodists because Calvinist predestination was underplayed, while their emphasis on the Bible and Jesus was not alien to the Quaker tradition.

Evangelicalism flourished best when there was an enemy to identify, and the danger in the 1820's came from free thinkers, rationalists, and Unitarians. In Massachusetts the old Puritan doctrines had gradually been undermined by rational religion. When the evangelical Congregationalists discovered the danger and sought to regain

---

16. Lois Banner, "Religious Benevolence as Social Control," *Journal of American History,* LX (June, 1973), pp. 23–41; David Bowen, Quaker Orthodoxy and Jacksonian Democracy, M.A. thesis, Swarthmore College, 1968.

Years of Crisis and Separation

control of the churches, the Unitarians, aided by court decisions on who could vote in parish matters, took control of most of the important Congregational churches around Boston. The battles between Unitarians and evangelicals occurred simultaneously with developing controversies in Philadelphia Yearly Meeting. Those Quakers who formed the Orthodox party thought they were fighting the same battle for Christianity going on in Boston.

The evangelicals accused their opponents of undervaluing Jesus Christ. The authoritative source of knowledge about Jesus was the Bible which proclaimed that Christ shed his blood as a sacrifice to satisfy the justice of God and thereby save men. No one could be saved without the blood of Christ, and the image of the suffering Jesus on the cross was central for orthodox piety. Christ was equal with God, a person in the trinity, born of the Virgin and without sin. The virgin birth, scriptural inerrancy, the trinity, a physical heaven (and hell) became important to evangelicals mainly as guarantees of the atonement. Orthodox Friends did not stress an emotional conversion experience, but did insist that certain revelation, obtained only in the Holy Scriptures, was authenticated by the experience of the Holy Spirit.[17] The main centers of strength of the Orthodox lay in the central institutions of Philadelphia Yearly Meeting: the Meeting for Sufferings, and the Select Meeting of Ministers and Elders. Philadelphia Yearly Meeting met only once a year and because of its size was a rather unwieldy body for operating mundane affairs. Authority tended to gravitate to the Meeting for Sufferings, but because the business of the Meeting for Sufferings was time consuming, and getting to Philadelphia was inconvenient, rural Quakers acquiesced in allowing the body to be dominated by the more prosperous Quakers who lived in the city and who could frequently attend sessions.

Effective power in the Yearly Meeting lay in committees chosen to draft epistles, decide difficult issues, and hear appeals. The most important yearly meeting official was the clerk. The clerk knew the agenda and appointed, with the consent of the main body, all committees. The clerk sat in front of the Arch Street Meeting House in the gallery surrounded by ministers and elders who had met just previous to the Yearly Meeting. Other preliminary decisions would be

---
17. *The Friend* (1827), I, pp. 38, 344, II, 25.

Friends in the Delaware Valley

made by the Meeting for Sufferings whose work would be reviewed and approved by the Yearly Meeting.

Modern theories of clerkship stress that the clerk is to be nonpartisan and to sum up the sense of the meeting. We do not know how accurately this describes the picture of the early nineteenth century, for Quakerism had a strong element of authoritarian domination by ministers, elders, and overseers. God had not yet become a democrat, and Friends had never believed that truth came by majority rule. Truth came by individually waiting upon the Lord, speaking his message, weightily considering spoken messages while anticipating agreement, and letting the clerk summarize the sense of the meeting. Those who were most devout, most attuned to communicating the will of God, in short, ministers and elders, had a predominant role in arriving at decisions. Theoretically there was equality of membership, but there was no equality in the distribution of spiritual gifts. Seventeenth century Puritan churches are described as a silent democracy with a speaking aristocracy (the ministers), and this portrait could be accurately applied to Philadelphia Yearly Meeting as well.

In addition to ministers, elders, and the clerk, the quarterly meetings appointed representatives who brought reports answering written queries summarizing the state of the meetings. These delegates had the right to speak in all matters of business. Supposedly all Friends had the right to attend and to speak, but in practice the clerk recognized only representatives, ministers, and elders. While all Friends met together for worship, men and women separated for business sessions. If Friends had confidence in the impartiality of the clerk, that he was willing to wait for the direction of Friends and able to be guided by the spirit of God rather than of party, then the Yearly Meeting could operate successfully. During the late 1820's a majority of Friends became convinced that the clerk had become a spokesman for a faction.

In opposition to the evangelicals were those Friends whom their opponents later termed "Hicksites," but who, like the "Orthodox," thought of themselves simply as Friends. The Orthodox are more easily identifiable because they had a common program: stamp out infidelity in the meeting. The Hicksite party formed in opposition to the Orthodox made little attempt to formulate their views in a systematic fashion.

Years of Crisis and Separation

The largest faction among Hicksites were traditional quietists who accepted the truth of Scripture, the divinity of Christ, and the resurrection. But the symbol they emphasized above all else was the indwelling Christ.[18] Truth in religion could be guaranteed by no external focus like the Bible, and theology was head learning of no great significance. Ministers preached under the direct leading of the spirit and the authenticity of their remarks was vouchsafed by their lives and by the response within the believers. The outward blood of Christ had once been shed, but people were now saved by the inward bearing of His cross. Bringing the light of Christ to bear upon a corrupt will was a long, difficult process and resulted not in a creed or belief but in a sanctified life. Justification and sanctification occurred together continuously through a Christian's witness. If all Hicksites had been quietists, there probably would have been no division, since there was a quietist majority also among the Orthodox.

Within the membership of those who became Hicksites were some—their numbers cannot be determined with any accuracy but they were a distinct minority—who also wanted to revitalize Friends but whose motivating impulses were formed by a more critical or rational approach to traditional Quakerism and Christian doctrines. Like the quietists, these members disliked creeds, but their opposition was based upon a newly emerging view of church history. The most distinguished church historian of the eighteenth century was a German Protestant named Johan Mosheim who viewed the development of doctrine as understandable only in terms of prevailing philosophical systems and political and social needs of the times. Mosheim attempted to treat the various "heresies" of the early church sympathetically and not to assume that a mysterious hidden revelation guaranteed the truth of the decisions of church councils. A Friend reading Mosheim's description of the developments of trinitarian doctrine could easily come to assume that the final formulation was an historical accident and not to be taken as equivalent in authority to the Scripture.[19] Elias Hicks is known to have owned at least two volumes of Mosheim's history; whether or not he read the

---
18. Elias Hicks, John Comly, Edward Hicks, and Jesse Kersey were quietists who wrote journals.
19. There were several English editions of Johan Mosheim, *An Ecclesiastical History, Ancient and Modern*. The Philadelphia edition of 1812 contained a refutation of the discussion of Friends signed by a clerk of Philadelphia Yearly Meeting.

complete work cannot now be established, but some of Hicks's emphases sound like Mosheim added to quietism. The Quakers who edited *The Berean,* a periodical first published in 1825 in Wilmington by David Ferris and James Gibbons, clearly had read Mosheim and absorbed some sense of his historical relativism.

A second source for Quaker liberalism (the term is here used for those criticizing evangelical doctrines on the basis of reason) came from rational Christianity and Scottish Common Sense philosophy. What these movements sought was a clear, reasonable, and benevolent deity who would appeal to man's intellect. Like the evangelicals, the rational Quakers wanted to ground religion on "facts" or "evidence."[20] Unlike the evangelicals, they insisted that mystery or emotion had little role in religion and only encouraged priestcraft, superstition, and despotism. Presbyterians were particularly suspect and the liberals accused them of using a benevolent empire of reform to create a religious establishment.

All the symbols most precious to the evangelicals seemed suspect to the liberals. Early Quakers had opposed the term trinity as unscriptural; liberal Quakers thought the whole concept absurd. God, a benevolent father, did not need a substitutionary atonement to satisfy his justice. In fact, just to attempt to conceive of an eternal being dividing himself into three persons and then deciding to sacrifice one part of himself to another made no sense. Liberals were quite willing to use a variety of methods of interpreting the Bible. Origen and the early church fathers had allegorized the creation story in Genesis. Why not do the same for other parts, for much of the Bible made no sense unless interpreted figuratively. Miracles proved the Bible was true, but how miracles occurred was not important, only the "fact" that they had happened.[21] Inward revelation—the term "inner light" now began to replace the inward light of Christ—served as a factual guarantee of the truth of not only Quakerism, but Christianity. If for evangelicals the blood of Christ was the prime symbol, for liberals it was freedom: freedom from dogma, from hierarchy, from superstition.

Most of the literature of the Orthodox-Hicksite separation was

---
20. *The Berean* (Wilmington, Del., 1826), I, p. 13.
21. *Ibid.,* pp. 27-28, 298; *The Friend,* I, p. 344, II, p. 382.

written from the perspective of either evangelicalism or liberalism. Both sides apparently forgot who their opponents actually were as they waged battles against imaginary and pernicious evil men.

Elias Hicks (1748–1830) appears almost incidental against the backdrop of intellectual and social division in the Society of Friends, but the controversies surrounding him brought on the schism and determined the way the two new groups would develop. Hicks's intellectual development is difficult to trace because he published no theological works until late in his career when his sermons, taken down in shorthand and published without his permission, appeared. Hicks's *Journal,* based upon travel diaries, was printed after the separation and after his death.

The character of Hicks's thought is best described as that of an extreme quietist or spiritualist tinged with rationalism. Formal theology he saw as head learning designed to foster pride and destroy godliness. Heaven, hell, and the atonement expressed spiritual truths and could be used to elucidate the inward journey of a true Christian. Hicks believed in the historic Christ, accepted the virgin birth (though recognizing that some accounts in the gospels and Paul made the doctrine suspect), the trinity, and the authority of Scripture. But Hicks insisted that such outward manifestations were of little significance. Inward revelation confirmed the truth of the Scripture, not vice versa.

During his travels as a minister, Hicks preached against outward knowledge of evangelical doctrines. When called to account for what his opponents termed excesses, he claimed immediate inspiration and insisted that he was in unity with the meetings where he spoke. If Orthodox accounts can be believed—neither Hicksite nor Orthodox should be automatically accepted—Hicks on one occasion compared the significance of the outward blood of Christ with the blood of a chicken.[22] As early as 1808 an English Quaker traveling with Hicks found his communications unsound, but many American Friends found Hicks an inspiring preacher and during his travels throughout America he attracted large crowds and kept the full support of his monthly, quarterly, and yearly meeting in New York.

---

22. Bliss Forbush, *Elias Hicks, Quaker Liberal* (New York, 1956); *The Friend,* I, pp. 204, 341; Elizabeth Robson quoted in Edwin P. Bronner, *The Other Branch: London Yearly Meeting and the Hicksites* (London, 1975), p. 7.

Friends in the Delaware Valley

Orthodox Friends were at a distinct disadvantage in the struggle over silencing Elias Hicks. Hicks, who was seventy-six in 1824, had spent most of his adult life in the service of Friends, and his quietist posture and eloquence reminded listeners, particularly in rural communities, of the kind of Quakerism with which they had grown up. His strictures on worldliness, against commercialism, and in favor of a strong antislavery position, including a boycott of materials produced by slave labor, found a sympathetic audience outside the city. Because Hicks insisted on the need for harmony and for the autonomy of the light, his opponents who claimed only to be defending Christianity seemed to many to be advocating thought control and denying freedom of worship. The liberals linked their cause with Elias Hicks and Green Street Monthly Meeting and represented the contest as between an emergent priest class, supported by visiting ministers from aristocratic England, against the efficacy of the inward light, holy living, and the autonomy of local meetings.

The Orthodox experienced frustration in dealing with what seemed a fundamental challenge to the faith. Elias Hicks appeared maddeningly evasive when the normally reliable disciplinary procedures failed to work. When elders and ministers attempted to declare the fundamentals of Christianity, they encountered opposition from country Friends both in the Meeting for Sufferings and the Yearly Meeting. Worse, they found themselves pilloried as pseudo-Presbyterians and aristocrats. In an age when rural-urban tensions were high and elitism was decried as unAmerican, these devout Philadelphia Quakers whose prosperity was a blessing from God and a tribute to their hard work were being defied by less educated, less affluent, less sophisticated farmers. Unable to silence Hicks or even gain an acknowlegment of error from him, and hampered by his sympathizers even within the city, the Orthodox decided to employ the agencies at their disposal: the support of English Friends, predominance in the Meeting for Sufferings and Ministers and Elders, and the clerkship of the Yearly Meeting.

From 1819, when Jonathan Evans,[23] a former clerk of the Yearly Meeting and elder of Philadelphia Monthly Meeting, first publicly manifested his disapproval of Elias Hicks, the controversy grew rap-

---

23. William Bacon Evans, *Jonathan Evans and His Time* (Boston, 1959) contains the best detailed description of the 1827 yearly meeting.

idly. Visiting ministers from England like William Forster, Anna Braithwaite, and Elizabeth Robson not only disagreed with Hicks but openly preached against him. A controversy in a religious journal from 1820 to 1823 between a Presbyterian minister and Benjamin Ferris, a Wilmington Friend, writing under the pseudonyms of Paul and Amicus, brought the charge that Quakers denied the essence of Christianity with a reply enunciating a liberal version of what Friends required.[24] The tone of the Quaker response irritated the evangelical members of the Meeting for Sufferings who proceeded to compile a list of extracts from the writings of early Friends to prove that Quakers were orthodox Christians. The extracts were strongly evangelical and the supporters of Hicks and "Amicus" objected to the publication of any such "creed" with the official imprimatur of the Meeting for Sufferings. The maneuverings and charges of both sides which climaxed in a debate in the Yearly Meeting showed the depth of distrust and animosity.[25]

In the city of Philadelphia all monthly meetings except Green Street opposed Hicks, but Green Street's members welcomed Hicks who continued to travel in the ministry and came to visit in 1822 and again in 1826. When Hicks attended different meetings and spoke, opponents testified against him and disruptions occurred. Convinced that Hicks did not believe that Jesus Christ was the only begotten Son of God, the elders attempted to forbid him from preaching. The Friends of Green Street, incensed that several of its elders cooperated in this venture, attempted to discipline one elder who sided with the Orthodox and to replace others. On appeal, Philadelphia Quarterly Meeting overruled Green Street Monthly Meeting and reinstated the elders, denying that local meetings had authority to replace elders who had done no wrong. When Green Street defied the Quarterly Meeting, the Orthodox moved to lay down the meeting; in response, Green Street attempted to secede from Philadelphia Quarter and to join Abington Quarter. Mt. Holly Meeting in New Jersey transferred to Bucks Quarter in an attempt to leave the Orthodox dominated Burlington Quarterly Meeting.[26] The issue now became con-

---

24. *Letters of Paul and Amicus, Originally Published in the Christian Repository* (Wilmington, 1823).
25. Forbush, *Elias Hicks,* pp. 218–220.
26. *The Friend,* I, p. 174; Samuel Janney, *History of the Religious Society of Friends* (Philadelphia, 1867), IV, pp. 236–245, 273–274.

stitutional: did effective power rest with a monthly meeting or a quarterly meeting? That question would be dealt with in the 1827 Yearly Meeting.

Since the 1760's in both England and America the yearly meetings had appointed committees to visit local meetings and persuade the members to achieve uniformity in disciplinary practices. In 1827, in spite of the opposition of some Friends, the Ministers and Elders decided to ask the Yearly Meeting for a visiting committee to investigate the doctrine of members.[27] The Hicksites saw the evangelicals as aiming at silencing and disowning all those who opposed them.

Samuel Bettle, Clerk of Yearly Meeting, testifying in court after the schism, declared that he was accustomed to recognize to speak and to weightily consider only the advice of reliable Friends, i.e. the Orthodox.[28] In an attempt to safeguard their interests, the anti-evangelicals appointed double the normal number of representatives to the 1827 Yearly Meeting. Country Friends decided that protection of their freedom required the elevation of the assistant clerk, John Comly, to the clerkship. The disunity manifested in monthly, quarterly, and yearly meetings was so distressing to Comly that early in 1827 he began discussing with certain members whether in the event of continued bickering a peaceful solution might be a temporary withdrawal.

The Yearly Meeting's sessions opened with a deadlock over continuing Bettle or nominating Comly as clerk, but after a protracted dispute Bettle continued to serve since he had previously been clerk. Sympathizers of Comly began holding special sessions in the evening at Green Street to debate appropriate action. When, in the last session, without general approval, a special committee composed of evangelicals was appointed to visit each meeting to test the soundness of the membership, separation became inevitable.

If relationships between Friends were unedifying before the Yearly Meeting, that bickering came to seem almost harmonious in the events that followed. Both groups termed themselves Philadelphia Yearly Meeting and claimed to be the only authentic Quakers; each

---

27. Philadelphia Yearly Meeting of Ministers and Elders, Minutes, 1827, p. 161; PYM, Minutes, 1827, pp. 412–429.
28. Jeremiah Foster, ed., *An Authentic Report of the Testimony in a Cause at Issue* (Philadelphia, 1831), I, pp. 36, 82.

accused the other of assorted despicable practices. The Orthodox position was that the Hicksites were neither Quakers nor Christians, and, since originally all Quakers had been Christians, all property should belong to them as legitimate heirs of the first Friends. Compromise and division of property with heretics would betray Christianity and the separatists should be disowned as soon as possible. Marriage between an Orthodox and a Hicksite was joining a Christian and an infidel. The Hicksite position was that property should be divided, probably on numerical lines, a position which had many advantages for the more numerous Hicksites. What about burial grounds, Westtown School, meeting records, etc.? People who had worshipped in a meeting house all their lives now were threatened with eviction. The burial ground in Philadelphia, formerly used by all monthly meetings, was now to be closed to schismatics who would not recognize the authority of the Orthodox Quarterly Meeting. When the Hicksites broke down part of the wall to build a new gate to give themselves access, the Orthodox had the perpetrators arrested and charged with trespass. In virtually all meetings there were scenes in which either Hicksite or Orthodox took control, and, while both sides proclaimed peace and love, they acted with passion and cunning. In case anyone should forget the heinousness of the other side, *The Friend* (Orthodox), *The Friend: Advocate of Truth* (Hicksite), and *The Berean* devoted their pages to proving their party's loving forbearance and dissecting the apostasy of their opponents. Personal vituperation, disownments, court trials, divisions of meeting houses with the new Orthodox structure just across the road from the Hicksite, expulsion of Hicksite pupils from Orthodox schools, and an outpouring of tracts kept the animosity burning for years.

As if the separation in Philadelphia were not serious enough, both sides attempted to gain vindication from other yearly meetings, and so managed to precipitate divisions in Baltimore, New York, Ohio, and Indiana. Correspondents in each yearly meeting sent to their side's publications inflammatory accounts of the separations. Since the editors, if no one else, read the opposing papers, they could fill pages by accusing their opposing counterparts of distorting the facts. Believing that their control of Friends' institutions required that only true Quakers remain on boards of directors, Orthodox Friends proceeded to disown all adult Hicksites. Unwilling to consign Hicksite children to perdition at once, the Orthodox waited until the children became of age and then visited them before disownment.[29] The Hicksites responded by disowning the Orthodox who were disowning them.

A trial in New Jersey over a trust fund showed a complete breakdown in communication between both sides. The Orthodox thought the division was over theology and insisted upon discussing fundamental beliefs; the Hicksites thought the division was over church government and, while refusing any discussions of theology, talked about the rights of members. They disagreed over the power and permanence of elders, the power of quarterly and yearly meetings over monthly meetings, the amount of equality within meetings, and whether creeds were only statements of personal beliefs or were binding upon all members.[30]

Only in the city of Philadelphia did the Orthodox have a clear cut majority. Elsewhere the Hicksites were dominant and they claimed that twenty-eight of the original monthly meetings remained Hicksite whereas only three became Orthodox. Burlington, Haddonfield, and Westtown had strong Orthodox constituencies. Believing that the number of adherents had some relationship to the value of their positions, both sides issued statements, supposedly based upon reports from observers or monthly meetings, but the statistics are not easily reconciled. Some Friends may just have stopped attending meeting in disgust and both sides claimed them. A convenient approximation is that Hicksites had 70 percent and the Orthodox 30 percent of the members.[31] Orthodox Friends kept control of Arch Street Meeting House, Westtown School, and Friends Asylum. Since most of the money to build these institutions had come from city Friends, this division of property had a kind of justice, but Hicksite Friends who had contributed to any of these institutions felt

---

29. PYM (O), Minutes, p. 278; *Friends Intelligencer,* I, p. 397.
30. Foster, *An Authentic Report,* I, pp. 21–22, 34–35, 50, 61.
31. *The Berean,* I, pp. 176–177 claimed a division of 15,699 to 4, 256; in court the Hicksite census claimed 18,485, Orthodox 7,344, neutral 429. Thomas Evans in six quarters found 1,979 more Orthodox and 1,332 fewer Hicksites. The Hicksite figures showed that the division in Philadelphia was approximately equal; the Orthodox claimed a majority of 2,926 to 1,461 adults. Two-thirds of the ministers and elders became Orthodox. Janney, *History of the Religious Society of Friends,* IV, p. 287; Foster, *An Authentic Report,* II, p. 495; *The Friend,* II, pp. 141–143, 204; *The Berean,* I, pp. 145–146.

Years of Crisis and Separation

aggrieved. Orthodox Friends felt bitter because the Hicksites took possession of meeting houses where some evangelicals worshipped.

Elsewhere Hicksites became the overwhelming party in Baltimore and New York Yearly Meetings; New England and North Carolina had virtually no Hicksites and no serious divisions. Ohio Yearly Meeting, where the numbers were nearly equal, had a separation where the bitterness and battles were, if possible, more acrimonious than in Philadelphia.

Within a year both Philadelphia groups had reorganized and during the same month held separate yearly meetings. The Orthodox met at Arch Street; the Hicksites at Green Street and the newly built Cherry Street meeting. Since these facilities were too small, planning began for a new centralized location and, in 1855, the new meeting house at Fifteenth and Race Streets was completed. Both yearly meetings attempted to keep communications with London Yearly Meeting and sent epistles expecting to have them answered. London remained strongly evangelical and English ministers had helped precipitate the schism by their stringent views. The predictable result was that London communicated only with evangelical yearly meetings and refused to recognize the Hicksites as Quakers. Orthodox relations with Presbyterians and Methodists also improved, since evangelicals in many denominations sympathized with the battle against infidelity. Naturally, the Hicksites found sympathy among Unitarians and Transcendentalists.

Reorganization

After the initial controversy had subsided, a main Orthodox principle became preventing a repetition of the Hicksite schism. The first requisite was a clear statement of what Friends believed, and a general gathering of representatives from Orthodox Yearly Meetings took place in Baltimore. The fruit of the meeting was *The Testimony of the Society of Friends in the Continent of North America* (1831), a tract of 30 pages, which carried the imprimatur of eight Orthodox Yearly Meetings. This was followed by the reprinting of works of early Quakers. In 1837 William and Thomas Evans issued the first volume of *The Friends Library: Comprising Journals, Doctrinal*

Treatises, and Other Writings. The Library spanned fourteen volumes, each with nearly 500 pages of small print. In 1831 the complete works of Robert Barclay were reissued and George Fox's Barbados Declaration of Faith. *The Friend,* issued monthly, provided a vehicle through which all the Orthodox could receive news, reviews of books, and accounts of Quakers. Equally determined to claim the vindication of early Friends, the Hicksites created a ten volume *Friends Miscellany* beginning in 1831, published the *Works* of George Fox in eight volumes, and reprinted William Penn's *The Sandy Foundation Shaken.*

The second phase of the Orthodox revitalization was the elimination of any ambiguity in discipline which had been cited by the Hicksites as justification. The result was a much more tightly centralized and hierarchical structure with power explicitly vested in the Yearly Meeting and the Meeting for Sufferings. In revisions of the discipline in 1832 and 1834 Philadelphia Yearly Meeting adopted new procedures for laying down meetings, disowning or disciplining ministers and members even if the monthly meeting to which they belonged did nothing, and restricting the freedom of ministers to travel and visit.[32] In matters of discipline, procedures were clarified. When a quarterly meeting conducted an arbitration between a member and the monthly meeting, the number of persons nominated to serve would be from eight to twelve and neither the monthly meeting nor the plaintiff could reject more than three.[33]

The power to disown over matters of belief was made explicit. Any member could be disciplined and disowned for "printing, selling, or distributing books or papers which tend to the denial or laying waste a belief in the divinity, mediation, and atonement of our Lord and Savior Jesus Christ, the immediate influence of the Holy Spirit, or the authenticity and divine inspiration of the Holy Scriptures."[34] Anyone who denied the propitiatory sacrifice of Christ "without us" was guilty of blasphemy.

Believing that the schism came because of widespread ignorance of Scripture, in 1829 a group of prominent Friends, including members

---

32. PYM (O), *Rules of Discipline* (Philadelphia, 1834), pp. 5, 48–49, 62, 88, 98–100.
33. *Ibid.,* p. 10.
34. *Ibid.,* p. 25.

Years of Crisis and Separation

from several yearly meetings, issued an appeal for a Bible Association of Friends in America whose main function was to print and distribute Bibles at little or no cost to Friends schools and families. The headquarters was in Philadelphia and with its auxiliary branches (by 1831 there were twenty-one) distributed Bibles throughout America. The Society estimated that during the first five years of existence it printed and distributed nearly ten thousand Bibles and eight thousand New Testaments.[35]

If the Orthodox assumed that belief in the inerrancy of the Bible and stipulations in the discipline would guarantee harmony, they were disappointed. Joseph John Gurney (1788-1848)[36] came to symbolize a basic disagreement between quietists and evangelicals which divided Orthodox yearly meetings throughout America. Gurney, a brother of Elizabeth Fry, was a wealthy English banker, a well educated theologian, a humanitarian interested in prison reform and antislavery, a prominent minister, and a cosmopolitan. He associated openly with Anglicans and Presbyterians and made an effort to relate Quakerism to evangelical Christianity without automatically assuming that on any matter of disagreement early Friends were in the right. In England Gurney's publications and preaching which emphasized Bible study, the vicarious atonement, and education brought opposition from quietists like Sarah Grubb and Thomas Shillitoe. Shillitoe, fondly remembered in America for his strong opposition to Elias Hicks, warned on his deathbed against Gurney's growing influence.[37] When in 1837 Gurney applied for a minute from the London Yearly Meeting of Ministers and Elders to visit America, an acrimonious debate occurred before permission was granted. Americans knew of the tensions in English Quakerism between extreme evangelicals (termed Beaconites) who claimed to be followers of Gurney and who left London Yearly Meeting in 1835, the

---

35. Edwin B. Bronner, *Sharing the Scriptures: The Bible Association of Friends in America* (Philadelphia, 1979).
36. David Swift, *Joseph John Gurney: Banker, Reformer, and Quaker* (Middletown, Conn., 1962). Gurney's works were well-known and controversial in America before the Hicksite separation, including *A Letter to a Friend on the Authority, Purpose, and Effects, of Christianity, and especially on the Doctrine of Redemption* (1824) and *Observations on the Religious Peculiarities of the Society of Friends* (1824).
37. William Hodgson, *Society of Friends in the Nineteenth Century* (Philadelphia, 1875), I, pp. 311-314.

moderate evangelicals who stayed in the meeting, and the quietists.[38] All these groups believed in the truth of Scriptures, all opposed Hicksite rationalism, and all believed in silent meetings. The issues were the role of outward helps in spiritual matters and the relationship of the Inward Light to the historic events of Jesus' life.

John Wilbur (1774-1856) of Kingston, Rhode Island, visited England in the ministry between 1831 and 1833 and his published letters showed his opposition to Gurney. Joseph John Gurney's trip to America in 1837 appeared to be an extraordinary success, but in New England Wilbur sought an interview in which he attempted to convince Gurney of his errors. Wilbur's preaching against Gurney prompted New England Yearly Meeting to ask South Kingston Monthly Meeting to discipline him; when the meeting refused, the Yearly Meeting abolished the monthly meeting and disowned Wilbur. The result was a separation with about five hundred supporters of Wilbur forming a yearly meeting of their own.

Both New England Yearly Meetings sought recognition from Ohio and Philadelphia Yearly Meetings by sending epistles. Officially receiving and reading an epistle at the beginning of a yearly meeting and then sending back an epistle in answer was tantamount to endorsement of that meeting. From 1845 to 1854 Ohio Yearly Meeting equivocated, either reading both New England epistles or neither. The clerk of the men's meeting was a Wilburite, and the clerk of the women's meeting a Gurneyite; neither could be replaced. Ohio was faced with a decision whether to accept the credentials of visiting ministers from both Yearly Meetings.[39] Finally, in 1854, the Yearly Meeting divided.

Philadelphia Orthodox now had to make a decision not just about the splinter group in New England but a major split in Ohio. Baltimore, New York, and London recognized the Gurneyite bodies in all places. At first Philadelphia recognized the larger (Gurneyite) body in New England, but appointed a committee to investigate the events. In 1850 Philadelphia recognized the Friends of both New

---

38. *Ibid.,* I, pp. 227-291. Gurney attempted to mediate the Beaconite controversy in England. Isichei, *Victorian Quakers,* pp. 44-53.
39. *Ibid.,* pp. 292-349, II, 112-166. The epistles of Philadelphia Yearly Meeting in 1838 and 1839 contain a strong Wilburite emphasis.

Years of Crisis and Separation

England meetings as Quakers and the committee labeled the treatment of John Wilbur by New England Yearly Meeting (Gurneyite) as illegal. The attempt at mediation of the conflict was unsuccessful.

Probably a majority of Philadelphia Orthodox favored recognition of the smaller Wilburite bodies, but it was clear that endorsing either party in Ohio or New England would occasion a major schism. Samuel Bettle, who had been clerk at the time of the Hicksite controversy, had concluded that that division was a tragedy and not worth the bitterness it had engendered. He wished at all cost to avoid a repetition, even at the cost of consistency. In 1855 Philadelphia suspended communication with New England Yearly Meeting (Gurneyite), though giving no reason for the action. Finally, in 1857 Philadelphia decided to have no correspondence with anyone.[40] Isolation might preserve a semblance of unity.

The two factions in Philadelphia Yearly Meeting (Orthodox) endured. Quietist strength in the city centered at Arch Street Monthly Meeting and in the Tract Society. Their publication was *The Friend*. Gurneyite members attended Twelfth Street Meeting and supported Haverford College. Their publication was *The Friends Review,* begun in 1847. Official yearly meeting sessions brought both groups together for worship and for cooperation on a limited range of traditional activities, Westtown School, and the Indian Committee. In the evening the Wilburites went home and the evangelicals met at Twelfth Street to discuss their distinctive concerns. The reaction of many Quakers was exemplified by Thomas Pym Cope who stopped attending all meetings for business because he did not like the acrimony.[41]

If in the 1830's the Orthodox had a quietist majority and evangelical minority, the Hicksites had a quietist majority and a liberal minority. But the Hicksites had an easier time of staying together because they believed in decentralized authority and were unwilling to disown over matters of doctrine. They did not, however, endorse a complete subjectivism. They believed that "God alone is the sover-

---
40. PYM (O), Minutes, 1857, p. 174.
41. Harrison, ed. *Philadelphia Merchant,* pp. 402, 417, 420–21, 430.

eign Lord of Conscience," but their discipline required disownment for denying the divinity of Christ or the "authenticity" of Scripture.[42] The Yearly Meeting made only slight changes in the discipline. To emphasize its subordinate position, the Meeting for Sufferings, when reestablished, was called the Representative Meeting and its power over publications diluted. Monthly meetings received the right to change elders and were told to review the appointment of all elders at fixed intervals. Friends disowned by the Orthodox could be admitted to Hicksite meetings on their own request.[43]

The liberal faction saw the division as an opportunity to reconsider certain disciplinary practices, but the quietist majority resisted basic changes of any kind, feeling that preservation of Quaker traditions was the surest method of proving continuity between George Fox and the nineteenth century. For example, in 1840 Friends were ready to stop endorsing traveling ministers' certificates, but then a few objections resulted in a continuation of the practice. That same year a long debate began over revision of the marriage discipline to allow a marriage between a Friend and a "non-member professor," probably an Orthodox member, if no violation of Quaker principles was involved (i.e. no hireling minister and the bride was not pregnant). After several years of being turned down and then reworded, the proposal was accepted. While the tone of the marriage discipline was more tolerant of other faiths, the Hicksites continued to disown over marriage out of unity. They debated whether any disciplinary offense which did not lead to disownment should be recorded. Friends had long borne a testimony against any kind of marker in a burial ground. Philadelphia Quarterly Meeting suggested in 1850 that, although extravagance in burial customs was improper, perhaps a 6 by 12 by 30 inch marker could be permitted. The matter was debated at the Yearly Meeting six times before the change was allowed.[44] To show their continuity with the past, the Orthodox meeting refused to allow this deviation.

The Hicksite Yearly Meeting warned against taking part in non-

---

42. PYM (H), Minutes, 1827, p. 4; *Rules of Discipline of the Yearly Meeting* (Philadelphia, 1831), p. 23.
43. PYM (H), Minutes, 1829, p. 47, 50–51; 1836, p. 159; *Discipline* (1831), p. 26.
44. PYM (H), Minutes, 1840–1855, pp. 52, 43–44, 91, 101, 293; 1847, pp. 147, 155; 1849, pp. 168, 172; 1852, pp. 239–240.

Years of Crisis and Separation

Quaker activities in 1838, against participation in political strife in 1840, and insisted that employing a chaplain at the state penitentiary violated the separation of church and state. Lucretia Mott was clerk of the Women's Yearly Meeting, but the meeting made no official move in favor of the women's movement. Rather, when in 1839 a proposal was made to grant women Friends equality with the males in the "formation, adoption, and administration" of discipline, the issue was referred to committee and never mentioned again.[45]

A reform movement associated with the Hicksites, though drawing members and ideas from others, was the Progressive or Congregational Friends of Longwood.[46] Established in the 1840's for religious, social, and educational changes, Longwood Friends advocated a religion of humanity which stressed the inherent goodness and perfectibility of man. They identified with what they saw as the spirit of Fox and Woolman and, like Emerson, preached an inner light which linked humanity to nature and God. Longwood Friends disliked dogma and their real commitment was to reform. Their causes exemplify antebellum crusades: Garrison-type abolitionism, temperance, women's rights, opposition to capital punishment, prison reform, homestead legislation, pacifism, Indian rights, economic regulation, and practical and co-educational schooling. The Progressive Friends originated out of meetings in Western Quarter but the relationship between Kennett and Marlborough Meetings and the reformers was often not harmonious. Speakers like Theodore Parker, Emerson, and Garrison were far too radical for most Hicksites. James and Lucretia Mott attended but never joined the Progressive Friends. Other Hicksites like William Logan Fisher and Thomas Garrett also came, but, unlike those in Ohio and New York, most Philadelphia Hicksites stayed at a distance from such heretical ideas. Edward Hicks, famous for paintings of the Peaceable Kingdom, was a quietist Hicksite who saw in the liberalism of the Motts and the Progressive Friends the destruction of traditional Quakerism. Hicks, who had strongly opposed the Orthodox, before his death in 1849 saw what he termed the essential similarity between the quietists in the two yearly meetings. In 1846 at the time of the death

---
45. *Ibid.*, 1839, pp. 21, 26, 29.
46. Albert J. Wahl, The Congregational or Progressive Friends in the Pre-Civil War Reform Movement, Ed.D. diss., Temple University, 1951.

of Jonathan Evans, the elder who so strongly opposed Elias Hicks, Edward Hicks wrote: "should I ever be permitted to enter the abodes of the ransomed and redeemed of the Lord, I shall hope to see the angelic spirit of dear Jonathan Evans and Elias Hicks, clothed in white raiment, with palms of victory in their hands, united forever."[47]

Education

Quaker educational ideals changed very little between 1780 and 1860. The first requisite for education remained that it be "guarded," for Friends assumed that their distinctive testimonies on speech, dress, peace, and marriage could not survive easily in an atmosphere of competition. Removal from the world's attractions into a controlled environment guaranteed not only the attainment of that modicum of practical schooling which Quakers regarded as essential, but restricted the emphasis upon too much head knowledge. The difficulty which plagued American Quaker efforts from the 1680's was finding enough Quaker pupils to attend schools taught by teachers of approved piety and knowledge. Even in Philadelphia Yearly Meeting there were many areas where Quakers became or remained small minorities and parents had to choose between no education or sending their children to schools with a mixed clientele.[48]

In England where similar problems existed a number of wealthy Quakers bought a former foundling hospital and created a new kind of institution, a yearly meeting owned and operated boarding school. Ackworth School opened in 1779 and became an instant success attracting mostly poorer classes at first but later expanding its curriculum and clientele. An Ackworth-type school offered to Philadelphia Friends the solution to several problems: a boarding school could be select, i.e., restricted to Quakers, and located in a rural area away from corrupting influences. Financed by wealthy Friends throughout the Yearly Meeting, it could provide a place for Friends to send their children, and could produce in time an adequate supply

---

47. *Memoirs of Edward Hicks* (Philadelphia, 1851), p. 131.
48. J. William Frost, *The Quaker Family in Colonial America* (New York, 1973), chapters 4–5.

Years of Crisis and Separation

of teachers for itself and other Friends' schools. The result was Westtown School, opened in 1799 as an institution created, owned, and supervised by Philadelphia Yearly Meeting.

The Westtown Committee reported annually to the Yearly Meeting, and changes in curriculum and practices show much about Quaker attitudes towards the wider culture.[49] The School was coeducational, geared to serving students beyond the elementary stage who entered at age nine or ten and left at fifteen to sixteen. While the school was thought of as a family, strict segregation between the sexes and harassment with petty rules were the norm. Girls, for example, were forbidden to walk in the woods or pick up fallen apples from trees. The plain style of dress was strictly enforced and clothes were altered at parents' expense if deemed unsuitably fashionable. In 1844 the education committee was greatly distressed that some students, evidently continuing habits learned at home, said "you" to a single person. There were no vacations at first and the school held classes as usual on Christmas. The curriculum was strong on practical subjects with boys being encouraged to keep gardens as leisure time activity. The school had better success in attracting girls than boys, perhaps because the teaching profession was feminized in the early nineteenth century. By 1824 Westtown had enrolled 1337 boys and 2145 girls, while 69 men and 181 women served as teachers for at least a short time.[50]

Before the separation the curriculum remained restricted. After 1830 in a general reorganization of the school, with lowered fees to attract enough students to enable Westtown to survive, instruction in the classics was instituted, set times for entrance and leaving decreed, vacations formalized, a period of examinations by the committee established, and instruction in works of theology including Barclay's catechism and tracts of J.G. Bevan became mandatory. Westtown slowly became more cosmopolitan in taste. Shakespeare quotations appeared in student literary society publications, and sports activities were carried out on a communal rather than individual basis with the most popular being football. Quaker strictness remained, but by the 1860's plays were performed and a few found surreptitious ways

---
49. Beginning in 1845 the Westtown reports were printed in *Extracts from the Minutes of Our Yearly Meeting*. Helen Hole, *Westtown Through the Years* (Westtown, Pa., 1942).
50. PYM, Minutes, 1824, p. 393.

of enjoying music. Westtown retained a strong Quaker flavor, but the curriculum was increasingly like that in a genteel Victorian school.

While Westtown engrossed most of the Yearly Meeting's attention and money before 1827, most Quaker children attended schools in their own localities. The queries show that Friends claimed to be educating their children in plainness and apprenticed them only to Friends. Occasionally the Yearly Meeting would provide a compilation of the number of schools taught by Friends or under the supervision of the monthly meeting. In 1807 there were sixty-four such schools, in 1817, seventy-two, and in 1822, seventy-nine.[51] When the schism occurred, both sides had their schools disrupted and the country Orthodox Friends had particular problems for they were widely scattered. Still, the Orthodox demonstrated an extremely strong commitment to schools in the next few years. In 1834 they reported forty-four schools under the care of meetings, nearly as many as the far more numerous Hicksites who recorded fifty-two schools in 1830 and fifty-five in 1839, but admitted that only forty-five were taught by members.[52]

Both Hicksite and Orthodox Meetings looked with skepticism on the rise of a public or common school system in Pennsylvania, in spite of the fact that the reformer most responsible for its creation was the Quaker Roberts Vaux. In 1843 in order to see whether the district schools really posed a severe threat, the Orthodox Yearly Meeting created a committee to investigate education. They reported a total of 1,714 school age children; of the 400 in the City of Philadelphia, three-fifths attended select schools, one-fifth were taught at non-select schools by Friends, and only one-fifth attended non-Quaker public or private schools. The 1,314 country Orthodox Friends' children included 308 in schools taught by members and supervised by the meeting, 127 at Westtown, 12 at select schools; 145 were in schools taught by Friends not under the care of a meeting, 204 were in family schools, 99 in common neighborhood schools and 202 in public district schools. In country schools, even those under supervision of the meetings, Quaker pupils formed a distinct minority of about one-third.

---

51. *Ibid.,* 1807, pp. 137, 154; 1817, p. 287; 1822, p. 367.
52. PYM (O), Minutes, 1834, p. 251; PYM (H), Minutes, 1832, pp. 88; 1833, pp. 106, 108.

Years of Crisis and Separation

Many rural schools were more like district schools than Quaker institutions because they were supported by a combination of private and public funds. Virtually all Orthodox schools had copies of the Bible, which was read regularly, and Lindley Murray's readers, but there were many other books which often contained principles abhorrent to Friends. Few schools had copies of Barclay's catechism or other Quaker books. Family schools were attended by the children of many Friends because 55 meetings in Pennsylvania containing 557 children had no schools under the care of the Society.[53]

In the 1830's the minutes of the Philadelphia Yearly Meeting (Hicksite) show no overriding concern with education and there was no attempt to create a replacement for Westtown. In the 1850's the Yearly Meeting appointed an education committee and received a series of detailed reports and recommendations. Of 2,509 eligible children, 1,067 were in schools either supervised by monthly meetings or taught by Friends, 48 were in boarding schools, and 1,048 attended district schools. The education committee voiced distress that over 40 percent were in non-Quaker institutions, warned that the survival of a distinctive form of religion was jeopardized, and advocated a series of reforms including creating endowments for scholarships and schools. The Yearly Meeting postponed action for several years. Finally, in 1854, after accepting the reform, the meeting piously declared, "The Importance of providing our children with suitable intellectual food, which is necessary for the growth and development of the mind, was acknowledged, but it was felt to be far more important that their moral and religious training would be attended to."[54] Obviously, little could be done through official meeting channels. The next step was a series of meetings by liberal Hicksites of several yearly meetings to create a boarding school and/or college. The outbreak of the Civil War postponed the creation of the institution until 1864.

In 1808 Haddonfield Quarterly Meeting asked whether a more advanced school than Westtown was needed. The Yearly Meeting, dominated by quietists who were suspicious of higher education, rejected creating an institution where gentlemen could "polish" their

---
53. PYM (O), Minutes, 1841, p. 350; 1842, p. 367; 1843, p. 403; 1844, pp. 404-405; *Extracts from the Minutes,* 1845, 1846.
54. PYM (H), Minutes, 1851, p. 219; 1854, pp. 268-271.

learning.[55] The separation changed matters, for many Orthodox Friends believed that a well-educated elite, sound in Quaker principles, could have kept many from being captivated by the sophistry of the Hicksites.

Many of those involved in the Tract and Bible Societies played prominent roles in the movement for a Friends Central School, including members from several yearly meetings. Philadelphia Yearly Meeting officially participated in some preliminary discussion, but the resulting organization was autonomous and legally incorporated as the Haverford School Association. The change in names may have been due to Hicksite influence in the Pennsylvania legislature which delayed granting a charter to such "false" Friends.[56]

There was no intention of creating a college, only an advanced school to keep Quaker boys from attending colleges dominated by other religious societies. The first articles advocating the new institution appeared in *The Friend* in 1830 and the school opened in 1833 with a faculty of four. The school offered an advanced curriculum in Latin, Greek, science, ancient history, and literature, but most students were young and enrolled in the preparatory department. The managers insisted upon strict control of the boys in dress, speech, and conduct through dozens of regulations. The authorities selected a few fit newspapers, barred fiction, and rigorously regulated conduct on First Days. Until the financial panic of 1837, the boarding school flourished with seventy-six students enrolled by 1836. By the 1840's enrollment dropped to the thirties, debts mounted, and in 1845 the managers suspended the school.

The institution was saved by the fund raising activities of alumni, but after it reopened in 1848 there were significant differences. The restriction allowing only Orthodox Friends to attend was dropped, an alumni association was created, some relaxation of rules occurred, and the caliber of the faculty and difficulty of the curriculum increased. Although Wilburite Friends were horrified, in 1856 the school petitioned the legislature for authority to grant degrees and became a college. By 1860 Haverford had sixty students, a distinguished faculty, and a bright future. The Wilburites in Phil-

---
55. PYM, Minutes, 1808, pp. 160, 163–164.
56. Philip C. Garrett, ed., *A History of Haverford College* (Philadelphia, 1892); Rufus Jones, *Haverford College* (New York, 1933).

adelphia Yearly Meeting remained suspicious, but the alliance between the Gurneyites and the College was firm.

Slavery

Compared with the situation in 1750 when slavery was legally established everywhere and rarely questioned, Quakers by the end of the Revolution could take considerable satisfaction in their role in awakening public opinion, and could point to several major accomplishments. Quakers, who had stood forth as having first freed their own slaves because of religious principles, then initiated an active program of philanthropy for blacks, now broadened their campaign to the political arena. In 1780 Pennsylvania passed a law providing for registration and eventual freedom for all slaves within its borders, a process so gradual that it was not completed until the 1840's. When asked to provide arguments for ending slavery to be used in England, the Philadelphia Meeting for Sufferings insisted that freedom hurt no class of Americans and occasioned the uplift of the Negro.[57]

In concern with other American yearly meetings, Philadelphia repeatedly petitioned Congress under the Articles of Confederation and the Constitution to end the slave trade and to endorse abolition. Stymied here by Southern opposition, more positive results came in a campaign in 1787 to tighten the loopholes in the Pennsylvania manumission statute and to persuade the legislatures of Pennsylvania, New Jersey, and Delaware to outlaw the fitting out of ships in their ports for the slave trade.[58]

The period from 1800 to 1860 was a time of searching for an adequate method of ending slavery in a way that jeopardized no Quaker testimonies. Friends preferred gradualism and moderate rhetoric without hatred to the slaveowner and with compassion for the black because these methods, they believed, had influenced Northerners to free their slaves. Petitions to Congress, thoughtful expositions on the

---

57. Edward Turner, *Negro in Pennsylvania* (Washington, 1912, reprinted 1969), pp. 85–87; Philadelphia Meeting for Sufferings, Minutes, 1786, pp. 13, 22–23; 1787, pp. 49, 63.
58. *Ibid.*, 1788, pp. 76, 79, 83, 92, 99, 110. The standard source on the entire period is Thomas Drake, *Quakers and Slavery* (New Haven, 1950).

causes for, results from, and sinfulness of slavery, and labors on behalf of freed blacks, combined with prayer and reliance upon the power of God seemed the best methods of softening the South. The difficulty was that moderation, sympathy with the Southerner's dilemmas, understanding racial prejudices, and working through the political system brought few positive results. Instead, slavery which in 1790 had been confined to a few states in the Southeast spread to the Mississippi and beyond, and even moderate remonstrances brought vituperation and threats from pro-slavery apologists.

The main official Quaker agency for working against slavery became the Meeting for Sufferings. Because many Quakers in the Pennsylvania Abolition Society (PAS) were also part of the Meeting for Sufferings, the work of the two organizations was closely coordinated. Both worked to provide adequate education for the free black population of Philadelphia; both worked against discriminatory legislation and attempted to locate jobs; both aided blacks in Pennsylvania who were in danger of being kidnapped and sold South; both lobbied with the Pennsylvania legislature; both supported the black community in establishing their own schools and churches. Manumission societies in the North and South had a substantial percentage of their membership from the Society of Friends, so close cooperation of the meetings and antislavery organizations was the norm. Quakers and members of the PAS had a mixed relationship with the American Colonization Society. So long as the Colonization Society worked against slavery, Quakers cooperated, but in the South the colonizers seemed more intent on exporting free blacks to Africa than in opposing servitude. Quakers thought that a refuge for blacks somewhere west of the Mississippi was a possible solution to the difficulty, but they thought the idea of sending blacks back to Africa chimerical. Friends and the PAS favored compensated emancipation and approved of a gradual process which would enable the slaves to make a transition to the responsibilities of freedom.[59]

Whether in providing information to English Friends in 1783, taking a census of the Philadelphia black community in 1843, or advo-

---

59. Wayne J. Eberley, The Pennsylvania Abolition Society, 1775-1830, Ph.D. Diss., Penn State University, 1973. If the articles in *The Friend, The Berean,* and *The Friends Intelligencer* were an accurate indicator, the Orthodox were far more sympathetic to colonization than the Hicksites.

cating easing restrictive laws, the Quaker position remained the same. Blacks, laboring under immense liabilities because of poverty and prejudice, in freedom evinced a moral character, intellectual capacity, and ability to work which was the equal of newly freed white indentured servants and others of the laboring poor. Quaker attitudes to the blacks remained tinged with a condescending paternalism and there was no effort to encourage membership in the meetings.

Moderation in confronting the evil of slavery seemed to some Friends to ignore the crucial issue: the economic advantage of slavery. Elias Hicks became a leading advocate of a total boycott of slave produced goods. The free produce movement was supported by Friends in the belief that if slavery became uneconomical, it would die. Even if slavery did not wither, Friends would stay morally free from supporting forced servitude. The Hicksites claimed that the origins of the animosity between Elias Hicks and the Philadelphia elders came from his strong antipathy to their lukewarm response to the free produce movement. The Orthodox denied this, and there were Orthodox Quakers who created free produce stores. The Hicksite Yearly Meeting recorded for years that some Friends interpreted the query against refraining from prize goods (merchandise from ships taken in war) as requiring free produce. The slaves had been captured in war and, therefore, their labor was prize goods. Neither yearly meeting ever endorsed this argument. As a method of demonstrating commitment to the cause of antislavery, the free produce movement had some successes but, because the boycott did not seem likely to end servitude, never attracted mass support.[60]

Slavery occasioned acute moral dilemmas for those wanting to obey the law yet unwilling to recognize the legal basis of the peculiar institution. Isaac Hopper mastered the intricacies of the laws concerning manumission, kidnapping, evidence, and slavery in order to serve as a defender of blacks preserving their freedom and those attempting to gain their freedom. Conservatives approved of the good Hopper did, but questioned whether Friends' support of fugitive slaves using legal technicalities was in keeping with Quaker atti-

---

60. Ruth Ketring Nuermberger, *The Free Produce Movement: A Quaker Protest against Slavery* (Durham, N.C., 1942); PYM (H), Minutes, 1837, p. 180.

Race Street Meetinghouse, Fifteenth and Race Streets, was built in 1856 by the Hicksites and used as their meetinghouse until 1955. Today it is part of Friends Center.

tudes towards honesty and law. In popular memory, Quakers conducted the underground railroad and smuggled thousands of blacks to freedom. In actuality, a few Friends like Thomas Garrett of Delaware lived up to the legend, but few Quakers had any direct contact with fugitive slaves.[61] The Christiana Riots of 1851, in which fugitive blacks used arms to defend their newly won freedom and killed a master, raised perplexing issues concerning the peace testimony. Undoubtedly most Friends would have quietly aided fugitives rather than return them to slavery. But a few Friends also recognized that helping a few runaways did not seriously weaken the slave system while complicating the difficult task of persuading the South of the evil of the institution.

In 1831, unhappy with the gentlemanly tactics espoused by manumission societies, William Lloyd Garrison left Quaker Benjamin Lundy's employ and founded *The Liberator*. The spectacular growth of the new form of antislavery attracted and repelled Friends. Garrison's followers preached immediate emancipation but riots resulted and the hostility of the South mounted. Quaker sympathizers created new abolition groups modeled on Garrison's principles, but Quaker opponents saw the new movement as fostering hatred for the South rather than meaningful reform and did not wish to be associated with violence. Before 1840 Philadelphia Yearly Meeting Orthodox and Hicksite Yearly Meetings in New York, Baltimore, and Philadelphia issued strong warnings against Friends joining in activities for good purposes with those who had not the proper religious sensitivities.[62]

Friends never came to an agreement on the proper tactics to use against slavery. Supposed temporizing of Indiana Yearly Meeting (Orthodox) occasioned a schism in 1842 over slavery; New York Yearly Meeting (Hicksite) confirmed the disownment of Isaac Hopper and other radicals because of their association with attacks upon Friends made by other abolitionists. In the 1840's radical Friends continued to associate with outside abolitionists; other Friends created the Free Produce Association; most continued the quiet protest against slavery by issuing pamphlets, supporting the free soil move-

---

61. Larry Gara, *The Liberty Line: Legend of the Underground Railroad* (Lexington, Ky., 1967).
62. PYM (O), Minutes, 1837, p. 309; PYM (H), Minutes, 1838, p. 16.

ment, opposing war with Mexico, and sympathizing with fugitives. The events of the 1850's, the fugitive slave law, bleeding Kansas, Dred Scott, and John Brown, paralyzed Friends who hated slavery and feared civil war. In 1860 in overwhelming numbers, Friends voted for the antislavery candidate, Abraham Lincoln, and hoped for peace and freedom.

Temperance

The movement of the Society of Friends from advocacy of moderation to temperance to total abstinence and the support of restrictive legislation on alcohol parallels the rise of the antislavery testimony. Both reforms began from a well established related concern, grew from a few Friends speaking out and writing tracts, continued slowly for a considerable time, eventually conquered in the meetings, and attempted to spread to the outside world by political means.

The initial temperance movement was started by antislavery Quakers, notably John Woolman and Anthony Benezet, who raised the moral issue as to whether Quakers should consume spirituous liquors which destroyed grain, brought exploitation of workers, and led to various kinds of excessive behavior as well as drunkenness. Woolman and Benezet advocated drinking water as more conducive to health and religion than drinking spirits.

Increasing worry about the effects of alcohol may have been occasioned by a technological revolution in the eighteenth century allowing the cost of producing gin and whiskey to drop drastically. Previously whiskey had been a rich man's beverage; now cheap gin, whiskey, and rum were widely available. On the American frontier making whiskey became a way of preserving the cash value of grain in an area which had poor transportation, and the prodigious drinking habits of the inhabitants made spirits a valuable medium of commerce. Initially for most Friends the testimony was not against the traditional beverages: wine for the rich and beer and cider for the poor. Rather, Quakers opposed converting grain into whiskey and gin.

Between 1784 and 1794 Friends were exhorted to use moderation in drinking within the family, forbidden to frequent taverns, and told to refrain from importing and vending liquor or selling grain to be

Years of Crisis and Separation

made into alcohol. In 1794 those who violated the testimony could be placed under a halfway disownment, unable to contribute to or be employed in any service of the church. Those who distilled or sold liquors were to be met with by committees. The monthly meetings were granted "liberty" to disown those who violated the testimony.[63] Beginning in 1797, there is a series of figures derived from answers to queries which were used by Friends to assess the progress of their temperance campaign. Early nineteenth century Quaker statistics need to be accepted with caution, because the numbers are derived from information given by quarterly meetings based upon figures from monthly meetings which depended upon cases reported from preparative meetings. A great many Friends still needed to be persuaded of the need for temperance regulations.

In 1797, with not all quarterly meetings responding, there were 9 Quaker distillers and 75 retailers; more accurate reporting by 1804 raised the number of sellers to 89; in 1806 there were 19 distillers and 104 wholesalers and the numbers varied little until 1823 when 102 dealt in spirituous beverages and 14 distilled.[64] The statistics indicate that little improvement was made, but they may be misleading. In 1823 Concord Quarterly Meeting suggested that the distinction between alcohol made from grain and from fruit was invalid. If several meetings interpreted the spirituous liquors to be products like cider, then there may have been a drop in the actual number of distillers from grain. Also, no information was reported as to the number of Friends who drank or what beverages they preferred. The reformers, particularly from rural meetings, were dissatisfied with the amount of change and for many years rural quarterly meetings would suggest additional measures only to find the Yearly Meeting divided. Before the schism the only additional step came in 1822 when Philadelphia recommended that Friends in a "private capacity" speak to those still using "the article as a common drink."[65] At the time of the schism, Samuel Bettle testified that only Ohio Yearly Meeting disowned for distillation of alcohol.[66] Clearly, whatever the Yearly Meeting might say, disownments over distilling even of whiskey remained rare.

---

63. Philadelphia Yearly Meeting, *Rules and Discipline* . . . (Philadelphia, 1797), pp. 86–89.
64. PYM, Minutes, 1797, p. 345; 1804, p. 91; 1806, p. 121; 1823, p. 379.
65. PYM, Minutes, 1822, p. 372; 1823, pp. 380, 383.
66. Foster, *An Authentic Report,* p. 91.

When Quakers began their temperance crusade, they were essentially alone, but they soon picked up two allies, the Methodist Church and the medical doctors. Until the 1830's the Quakers, Methodists, and a few physicians made only slight progress, but the Second Great Awakening spilled over into social concerns and one of the new causes was temperance. As the temperance movement spread, it divided into two conflicting parties. One group insisted on total abstinence from all alcoholic beverages, including cider and whiskey; the other advocated abstinence from gin and whiskey but moderation in the use of wine, cider, and other fermented drinks.

Many Orthodox believed in total abstinence, but their Yearly Meeting did not officially go beyond moderation and a restrictive definition of spirituous liquors. One cannot be certain whether practices of the members of the two yearly meetings diverged appreciably since only the Orthodox compiled information on those who drank or served liquor. After 1832, when the discipline required disownment of all who distilled alcohol, there are no records of any violators. The query asked about Friends' drinking or serving alcohol to guests or workmen, and those figures remain virtually unchanged until the Civil War. In 1839 the Orthodox found that sixty-nine Friends had not seen the evil in drink; in 1858 the number remained at seventy-one.[67] The discipline added no new regulations condemning drinking, but the thrust of the meeting's counsel was clearly against consumption, except for medical purposes. Still, at Westtown School in the 1860's at an officially sponsored picnic, wine was served to students, suggesting that the testimony remained only against whiskey and gin.[68]

The Hicksites ordered all who distilled spirituous liquors disowned and ended, officially at least, the distinction between wine and whiskey. In 1841 the meeting told members to abstain from renting property or furnishing any materials used in the manufacture or distribution of spirituous liquors. But, unlike the Orthodox, the Hicksites in 1832 advised Friends to avoid the use of "spirituous liquors *as a drink*" (italics in original).[69] Because the Yearly Meeting did not take

---

67. PYM (O), Minutes, 1839, p. 330; *Extracts from the Minutes,* 1858.
68. Hole, *Westtown Through the Years,* p. 197.
69. PYM (H), *Rules of Discipline* . . . (Philadelphia, 1856), p. 72.

a census of drinkers, the members' compliance remains unknown. The only numbers provided discuss distillation and selling. In 1842, the Hicksites began counting again and found no distillers but twenty-one or twenty-two continuing to sell liquor. After much labor, the numbers dropped to five in 1846 and two in 1856.[70] Many Hicksites included cider and wine within the definition of spirituous liquors, but no one was put under dealing for moderate drinking.

While Hicksites demanded great freedom in theological matters, their attitudes on temperance show little tolerance. The Progressive Friends of Longwood willingly sponsored lectures by clergymen from many denominations because conscience could not be coerced. But when the movement to restrict the sale of alcohol by voluntary consent seemed to be faltering, the Progessive Friends, including Lucretia Mott, actively lobbied for a state law enforcing prohibition. Neal Dow, a birthright Quaker from Maine, pioneered the legislative phase of the temperance movement and Maine went dry in 1851. The Progressive Friends were bitterly disappointed when in 1854 a referendum to make Pennsylvania dry was narrowly defeated.[71]

Conclusions

During the first half of the nineteenth century Philadelphia Friends pursued and expanded traditional concerns while earning a reputation for engaging in wide scale philanthropic activities. Individual members, often with the support of their meetings, made major contributions to pre-Civil War reforms: women's rights, public education, moral purity, the democratizing of politics, the creation of hospitals for the insane, and the establishing of penitentiaries. Antislavery became the major American crusade of the period, and the Quaker position against slavery remained constant and well-known. Friends experimented with many types of antislavery activities in their unsuccessful search for a method gradually to achieve emancipation.

---

70. PYM (H), Minutes, 1842, pp. 81, 83, 97, 135, 296.
71. Albert John Wahl, The Congregational or Progressive Friends in the Pre-Civil War Reform Movement, Ed.D. diss., Temple University, 1951, pp. 220-240.

Friends in the Delaware Valley

The meetings' school system in 1860, far larger and with better facilities and teachers than in 1780, attracted large numbers of pupils. Quakers continued to advocate a "guarded" education but expanded the curriculum, and the system encompassed a college, grammar and boarding schools, and institutions designed for girls, apprentices, children, orphans, and blacks.

The colonial testimony against drunkenness was transformed first into agitation against distilled spirits, then into advocacy of moderation or temperance, and eventually into total abstinence (except for medicine). A few Friends came to believe in legislated prohibition, a position not embraced by the nation until the first World War.

The central event between 1783 and 1860 was the separation of 1827. In retrospect, the schism did immense and irreparable harm to Philadelphia Yearly Meeting and diluted the impact of Friends upon American society. Virtually all Quakers continued to press for Indian rights, temperance, pacifism, and legal equality for blacks in South and North. The meetings' methods for changing people were love of neighbor, moral example, and appeals to reason. Against the rising tide of animosity between regions, Friends pleaded for moderation, gradual reform, and respect. Unfortunately, everyone could see that Quakers did not practice towards each other what they advocated for everyone else. For example, the approach of two separate delegations of Friends to the President and Congress on behalf of the dispossessed Cherokees meant that neither group could command as much moral, let alone political, weight. The meetings continued to address the world in pontifical tones, but often few outsiders paid attention.

A second result of the division was the removal of countervailing tendencies within segments of a divided Quakerism. A system of carefully balanced emphases—inward light and scripture, quietism and evangelism, individual initiative and group authority—built up over one hundred and fifty years, disappeared as Orthodox and Hicksites engaged in selective exegesis of biblical and Quaker sources. The results were simplified versions of the faith in which neither body could resist additional controversies. Some Hicksites flirted with Unitarianism, Transcendentalism, and Spiritualism, while some Orthodox espoused a narrow revivalism and others canonized an ossified quietism. Midwestern Quakers divided periodically throughout the nineteenth century because of such disagreements, but at least they enjoyed steady numerical growth through the century. Philadelphia Yearly Meeting managed to avoid these later separations, but it had its largest number of members before 1827, and, throughout the rest of the century while the metropolitan area increased spectacularly, dwindled to numerical insignificance.

A positive but unintended aftermath of the separation was the dawn of toleration within the Society of Friends. During the seventeenth century Friends fought for religious freedom in England and pioneered the creation of a colony without an established church, but toleration of error stopped at the boundaries of the meeting. A salient feature of Quaker apologetics throughout the colonial period was the amount of assurance, or, as their opponents saw it, spiritual arrogance. Friends possessed the gospel in its pure primitive form and any deviation was apostasy. Individuals in other religious bodies could be saved, but their salvation came in spite of and not because of their adherence to a false or inferior form of Christianity. Such confidence or rigidity could be easily maintained if virtually all members, or all influential ministers and elders, agreed on what constituted the truth. In the 1820's Philadelphia Friends made the traumatic discovery that the eighteenth century consensus had vanished and could not be refashioned. Ministerial authority, disownment, tract writing, vituperation and schism only brought to light additional disagreements. Ever so slowly, and most begrudgingly, Friends learned to live with, respect, and even love individuals within their meetings with whom they had profound differences. Such tolerance of diversity brought the birth of modern Quakerism.

I am indebted to Hugh Barbour, Edwin B. Bronner, Arthur Mekeel, P. Linwood Urban, and Patrick Henry for reading this paper and for their suggestions for improvements. "Beau" Weston wrote a term paper on the origins of the Quaker temperance movement which clarified many issues.

Friends in the Delaware Valley

Bibliographic Note

In comparison with earlier periods of Quaker history, the secondary literature on post-revolutionary Friends is sparse. The best books on the schism (Grubb, Doherty), slavery (Drake), and biographies are cited in footnotes. There are no recent monographs on such subjects as Indians, temperance, education, economics, organization of the meetings before or after the separation, and the role of Friends in the cultural life of the Delaware Valley. Sydney V. James, *A People Among Peoples: Quaker Benevolence in Eighteenth Century America* (1963) contains chapters on changes in various testimonies in the early nineteenth century. Peter Brock, *Pioneers of the Peaceable Kingdom* (1970) is the standard work on the peace witness in the War of 1812, the Mexican War, and the Civil War. Anthony Wallace, *Birth and Death of the Seneca* (1970) contains a perceptive analysis of one missionary endeavor. For an understanding of the period, nothing supersedes Lydia Maria Child, *Isaac T. Hopper: A True Life* (1854) and Edward Hicks, *Memoirs* (1851). Hopper was disowned because of his antislavery activities and Hicks' autobiography was not subjected to the Meetings' censorship. Both are delightful reading.

III A Time of Change: Philadelphia Yearly Meeting, 1861–1914

by EDWIN B. BRONNER

The Yearly Meeting at Arch Street met annually in mid-April, and the Race Street Friends in mid-May. The sessions usually lasted for five days, beginning on Second Day or Monday. The ministers and elders convened on the previous Saturday, and it was customary to have enlarged meetings for worship in the city meetings on First Day. Ten men served as presiding clerk of Arch Street during these decades, and seven of Race. However, only two of the Orthodox clerks remained in office more than five years, while four of the Hicksites served a minimum of eight years. The fact that William Evans, clerk of Arch Street in 1861 had been in the chair ever since 1831 may have influenced what Friends did after he retired.

Joseph Walton (1817–1898) served as clerk of Arch Street Yearly Meeting from 1881 to 1896, a period of sixteen years.[1] He had been in the furniture business in his early years, but turned more to Quaker concerns as he grew older. He served as editor of *The Friend* for some period, and also as clerk of the Meeting for Sufferings. A powerful leader of the Wilburite element, he is quoted as telling a Gurneyite that those in his wing "should cultivate a 'spirit of submis-

---

1. Biographical sketches of virtually all the persons mentioned in this chapter may be found in the "Dictionary of Quaker Biography" in typescript. These may be consulted in the Quaker Collection, Haverford College; or in the Library, Friends House, London.

siveness' to authority."[2] He was also a student of biology and geology. He should not be confused with Joseph S. Walton (1855-1912), prominent educator and later principal of George School, a minister in the other yearly meeting.

The other long term clerk of Arch Street Friends was Joseph Scattergood (c. 1808-1877), who served a dozen years from 1865 to 1876. A chemist and pharmacist in Philadelphia, he was active on the Westtown Committee for many years, was clerk of Meeting for Sufferings, and an elder in his meeting. Emmor Roberts (c. 1831-1908) was a farmer in the Moorestown area of New Jersey, and an active supporter of Swarthmore College, who served as clerk of Race Street Yearly Meeting from 1886 to 1900. His predecessor was Benjamin Foulke (1813-1888), clerk from 1873 to 1885, which meant that these two men filled the post for twenty-eight years. Foulke came from the rural area north of Philadelphia, but trained as a surveyor and helped to lay out some of the early railroads. He also had a good bit of knowledge about real estate law and orphan court affairs.

One immediate difference between the two Yearly Meetings in session was the presence of visiting ministers and other concerned Friends from other Hicksite yearly meetings at Race Street.[3] These men and women added an extra dimension to proceedings, especially to times of worship. In addition, Philadelphia received epistles from the other bodies in that branch of the Society, and asked a special committee to compose an epistle each year to send to the others. Arch Street Friends had severed all communication in 1857, and recognized no official visitors during the business sessions.[4] Nor did they send or receive epistles from other yearly meetings.

While there are not accurate membership figures available, it seems clear that Race Street was more than twice as large as Arch.

---
2. Philip S. Benjamin, *The Philadelphia Quakers in the Industrial Age, 1865-1920* (Phila., 1976), p. 18. This book was most helpful, and I have relied upon it in many ways. Hereafter: Benjamin, *Philadelphia Quakers*.

3. References to the two Yearly Meetings in session refer to the printed extracts. The year of reference is included in the text, but the actual page is not indicated in notes.
Extracts from the Minutes of Our Yearly Meeting Held in Philadelphia . . . (Phila., 1861-1914). The title changed from "Our" to "the" in 1864. (Orthodox).
Extracts from the Minutes of the Yearly Meeting held in Philadelphia . . . (Phila., 1861-1914). (Hicksite).

4. Traveling ministers were present on occasion from other yearly meetings, but they were not given official recognition.

A Time of Change

Philadelphia members made up a much larger proportion of the total membership in the smaller Yearly Meeting than in the other. Each Yearly Meeting continued to collect the quota to pay expenses on a proportional basis, and Philadelphia Quarterly Meeting (Arch) paid 35 percent of the whole, while in Race Street, the Philadelphia share was only 20 percent. Attendance at Yearly Meeting seemed to be approximately the same in each group; the larger proportion of city Friends in the smaller body may help to account for that. On the other hand, the quota for Arch Street was always larger than that for Race. In 1861 it was $1,500 for the former, and $600 for the latter. A decade later it was $1,500 for Race, but by that time was $4,000 for Arch. In 1885 Orthodox Friends collected $6,000, and the Hicksites just half of that amount.

The years covered by this essay fall rather neatly into two parts, especially in Race Street Yearly Meeting. Conditions remained virtually unchanged from 1861 to 1890, and then began to change rapidly in the last quarter century. In Arch Street the changes only became readily apparent in the new century, and were moving ahead vigorously in 1914 when this chapter comes to a close.

In the early years the pattern of business was quite similar in the two Yearly Meetings, with two days set aside each year to consider the answers to the queries. In each Yearly Meeting the week ended with a summary of the condition of the yearly meeting as a whole, and of the particular sessions just concluding. Sometimes these summaries were quite brief, and occasionally they were as long as five pages, depending upon the perceived needs of the times.

Beyond the similarities there were also obvious differences. One gains the impression of very tight control over the proceedings at Arch Street. Few questions came up from subordinate meetings, most issues were settled without long discussion, and no new ideas were introduced by visitors or incoming epistles. The only new committee created before 1890 was one to further the traditional concern for the education of the children in schools operated by Friends. The meeting was dedicated to preserving what had been, and the letters sent to local meetings urged members to be faithful to wear the plain dress, reminded men that they were to keep their hats on in meeting, and repeated earlier warnings about attending religious services conducted by others.

Friends in the Delaware Valley

However, at Race Street hardly a year went by when some new concept was not introduced, and more of these were adopted than rejected. In 1861 the Yearly Meeting of men accepted the principle that the women's Meeting was fully equal to theirs, and women were appointed to a number of committees during these decades, including Representative Meeting. After discussing the matter for several years, the Yearly Meeting agreed that members might marry non-members, as long as the spouse was sympathetic with Friends' principles. In 1866 the meeting expressed approval of the idea that efforts should be made to seek unity of all the different kinds of Friends. The body approved creation of a First Day School Committee during this period, and formed a new committee to work for the education of minorities, particularly blacks. It also had a committee on non-resident members, as well as ones on Indians, education and temperance. The Race Street Yearly Meeting sent more petitions to those in government about the need for reforms than Arch, and appeared to be more involved in the non-Quaker world than the parallel body.

Both Yearly Meetings, however, strongly supported the peace testimony during the Civil War, and devoted much time to exhorting members to be faithful in regard to temperance, by which they really meant total abstinence. Each Yearly Meeting made a concerted effort during the period to shore up the spiritual life and vitality of its members through the work of a special committee which visited individual members. Both bodies expressed opposition to the tendency toward a paid ministry which had come into existence in mid-western Gurneyite yearly meetings in the period.

Visiting Friends from overseas sometimes found it difficult to tell which group of Friends they were with, especially if they went into a meeting house without discovering ahead of time the affiliation of the body. Outsiders must have been even more confused about the two branches of the Society. If this was true of individual Quakers and local meetings, it was not yet apparent in yearly meeting sessions. The group which met annually at Race Street changed substantially in the 1890's, while Arch Street remained virtually unchanged, except that it was becoming smaller. The Orthodox Friends in the two quarterly meetings north of Philadelphia on the two sides of the Delaware, Bucks and Burlington, shrank in size until they asked to be combined in one quarterly meeting in 1898 called Burlington and

A Time of Change

Bucks. In 1904 Salem Quarterly Meeting was laid down, and the meetings joined Haddonfield, which graciously agreed to change its name to Haddonfield and Salem Quarterly Meeting. At the turn of the century, Race Street reported 11,500 members, and Arch, 4,300.

Friends Face a New Century

John Malin George (1802-1887), a member of Radnor Monthly Meeting, and the leading Friend in old Merion Preparative Meeting, left his entire estate to the Race Street Yearly Meeting for a boarding school. There had been numerous proposals that a yearly meeting school be created to parallel Westtown, but until the George bequest, Friends had been unable to unite on the idea. Plans were made very carefully during the next several years, and the school opened on land near Newtown in Bucks County in 1894. The new institution admitted non-Friends from the very beginning, and did not require that the teachers be Quakers, although only two of the original faculty were not. Race Street Yearly Meeting lavished much care and concern on this new enterprise. In the meantime, most of Westtown was rebuilt in the 1880's; the funds for this project, $332,000, had all been raised by 1889.

A new Committee on Philanthropic Labor was formed by Hicksite Friends in 1891, which paralleled the biennial conferences held by the yearly meetings sharing a common origin. Philanthropy was regarded as a desirable goal in the nineteenth century, and might be likened to present day committees on social concerns. Originally there were three subcommittees: on Indian work, temperance, and the education of blacks, referred to as "colored people." Later in the decade new subcommittees were named to further efforts for peace and arbitration, social purity, improvement of the conditions of working women, and against improper publications. In the twentieth century subcommittees were named to work for the equal rights of women and for prison reform. All of this activity led to discussions about whether the emphasis upon Philanthropic Labor diminished attachment to "religion," and provoked the rejoinder that Philanthropic Labor is religion.

The work with and for members of the Yearly Meeting gained a strong impetus from several capital funds given by Anna T. Jeanes (1822-1907) in the names of her brothers. One fund was to aid the education in Friends schools, a second was to aid the homes for aging Friends, and a third was for the building and repair of meeting houses. By the turn of the century a special Board of Trustees was named to handle all of the capital funds of the Yearly Meeting. The First Day School Committee was active in promoting religious education in the meetings and in some places where there were no Friends. Another group was named to correspond with distant members. In both of these efforts Philadelphians cooperated with Friends in the other Hicksite yearly meetings through biennial conferences. By 1910 there were so many committees active in the Yearly Meeting that a General Nominating Committee was created to propose names of persons to serve.

In 1900 the Friends who gathered in even numbered years to advance the various concerns of Quakers decided to form the Friends General Conference which would be guided by a Central Committee and incorporate the four separate gatherings (First-day School, Philanthropic Labor, Religion, and Education), which had long met at the same place. Philadelphia Yearly Meeting took a leading role in this loose organization for it was much larger than any of the other Hicksite bodies. Moving ahead in another direction, the Friends Central Bureau was created in 1911 to provide a year-round staff to serve the various committees of the Yearly Meeting. The first head of the Bureau was Jane P. Rushmore (1864-1958), who came to Pennsylvania from New York in 1884 to teach at London Grove. She soon took an active part in the Yearly Meeting, as well as the Friends General Conference, and was well equipped to fill this important new post.

There were some Friends who were concerned about all of these activities, and longed for more spiritual depth in the Yearly Meeting. One of these was Margaretta Walton (1829-1904), sister of the educator Joseph S. Walton, and aunt of George and Barnard Walton. She served as clerk of the women's Yearly Meeting for eight years, and travelled in the ministry among all of the groups that were to be in the Friends General Conference. On one occasion she wrote in her journal, "I don't think we once touched bottom and brought up real living stones. We touched Peace, Temperance, Philanthropical work,

Social Purity, etc."[5] She sometimes felt she was nearly alone in her concern, but it is clear from the exercises of each annual session that others shared the need to maintain a balance between spiritual concern and social action.

In the new century Arch Street began to show a greater awareness of the world around it, and to adopt various steps to bring about improvement. These changes came slowly, and were only claiming major attention at the end of the period covered by this essay. While the Gurneyite wing of the Yearly Meeting had been carrying on a wide range of activities in the social reform area through independent organizations, it was not until the committee to aid the Doukhobors was formed at the turn of the century that the Yearly Meeting began to accept responsibility in this sphere.[6]

The Meeting for Sufferings, which was renamed Representative Meeting in 1911, made a full report to Yearly Meeting early in each session. The larger body approved the various memorials and petitions to government authorities against gambling, capital punishment, etc., and issued pamphlets to advance the principles of peace. News of persecution of young Friends in Australia under a new compulsory military law aroused the sympathy and support of the Yearly Meeting. Even so, it was not ready to approve a proposal from Abington Quarterly Meeting that a new query be written which would ask whether members were showing proper concern about the social ills which afflicted humanity. A special committee named to consider the proposal approved of the idea, but rejected a query. However, in 1914, the Yearly Meeting sent a telegram to the President urging him to withdraw the army's punitive expedition from Mexico, asked local newspapers to tone down the war spirit which had surfaced, and adopted a resolution in favor of the Eighteenth Amendment, the so-called Prohibition Amendment.

The Orthodox Yearly Meeting Education Committee had employed a person in 1895 to give one day a week to visit schools in order to improve the schooling, and it slowly increased the time used to schedule conferences and provide assistance. In 1903, Anna

---
5. *Diaries and Correspondence of Margaretta Walton* . . . (n.p., 1962), p. 130.
6. Howard Brinton wrote about the yearly meeting as he remembered it from his youth. "Friends for 75 Years," *Bulletin* of Friends Historical Association (now *Quaker History*), 49 (1960), pp. 3–20.

Walton (1859-1954), who had served since the beginning, was given the title of Visiting Superintendent, which must have dismayed some Friends. The Race Street Committee began to hire an educational consultant in 1881. Arch Street Friends began to develop a concern for absent members in 1905, and sent an epistle to all such members. The other Yearly Meeting had been working on this concern for fifteen years. In 1913 the Orthodox agreed to form a committee to reach out to persons just outside the membership, in other words, to engage in outreach. When this step had been advocated by mid-western Friends thirty years earlier, *The Friend* had rejected the proposal, saying that there was too much likelihood it would be undertaken "in the will of man," rather than the will of God.[7]

The Yearly Meeting agreed to change the time of the annual gathering to coincide with the spring vacation of Westtown and the other Friends schools, in order to encourage wider participation. Race Street could not reach unity on following this example. On the other hand, it was only in 1914 that Arch Street agreed to admit women to Representative Meeting, something Race had done forty years earlier. The Orthodox body also revived the practice of recognizing and supporting Friends with a concern to travel in the ministry, and developed a new concern for Quakers in other parts of the world as they heard from such ministers.

Thus we see that there had been substantial changes in each yearly meeting since the beginning of the period covered by this essay. The quietist element seemed to be dominant in both yearly meetings at that time, and now the liberal forces held sway. London had gone through a similar change during these decades, although there the earlier influence had been more evangelical than quietistic. Many of the protestant denominations were struggling with the same issues, and the liberal element had been victorious in some churches. Midwestern Quakers had also gone through a good bit of turmoil in these decades. A new kind of evangelical spirit had triumphed over the quietistic wing, except in the conservative yearly meetings. The struggle with liberalism did not break forth in any large degree until the post World War I era in the West.

In the Race Street body the changes came at a steady pace, for as

---

7. *The Friend,* 54 (1881), May 7, pp. 311-312.

A Time of Change

new proposals were made there was more willingness to consider them than in the other Yearly Meeting. Hicksites prided themselves on being tolerant of the views of others, and placed a high premium on individual conscience. They were comfortable with the principle of "unity in diversity," and put it into practice.[8] As late as 1911 Arch Street supported the principle that each member should be willing to give up some part of his or her personal beliefs in order to reach a single unity. The faculty members at Swarthmore, and after it was founded, George School, had a liberating influence on the Yearly Meeting both directly, and indirectly through their students who took their places in the Yearly Meeting. The *Friends Intelligencer* was an influence for change, especially after it merged with the *Friends Journal* in 1885. Howard M. Jenkins (1842-1902) served as editor of both publications, and used them to introduce new ideas.

Within the fold of Arch Street the *American Friend* had a similar influence within the Gurneyite wing under the editorship of Rufus M. Jones (1863-1948). He took over the *Friends Review* in 1893 and in the following year combined it with the *Christian Worker,* published in Chicago, to form the *American Friend.* Jones, who was also on the faculty at Haverford College, had been brought to the Philadelphia area from Maine by President Isaac Sharpless (1848-1920) to share in the effort to bring about a change in Philadelphia Orthodox Friends. *The Friend,* organ of the Wilburite wing, changed its format and policy in 1913, after the *American Friend* was transferrred to Richmond, Indiana, and reflected the various viewpoints within the Yearly Meeting thereafter. This meant that the liberal ideas which were stirring in the Yearly Meeting began to enter the homes of its readers.

One of the young Friends whose influence was felt in the years 1913 and 1914 and thereafter, Francis R. Taylor (1884-1947), published an article a dozen years later in which he gave his interpretation of what brought about the change in the Yearly Meeting.[9] He gave a great deal of credit to Isaac Sharpless who changed Haverford

---

8. In 1909 the Race Street meeting house was the site of a conference on "Unity of the Spirit," held by the National Federation of Religious Liberals. The president of the organization was Henry W. Wilbur (1851-1914), and speakers included William I. Hull, Isaac Clothier, and Jesse H. Holmes. *The Unity of Spirit:* Proceedings . . . (Boston, [1909]).
9. "The Development of the Activities of the Yearly Meeting (1827-1927)," *The Friend,* 101 (1927), Oct. 13, pp. 191-193.

College from a small, parochial institution into an influential center of intellectual activity and social concern. He added that, at the same time, the Westtown alumni organized The Westtown Old Scholar's Association in order to bring about changes and a modernizing of that century old institution. Soon there were many evidences of cooperation between Haverford and Westtown, replacing the separation and even suspicion which had prevailed at an earlier time. Taylor credits Sharpless, a Westtown alumnus, with being a prime figure in closing the gap between Wilburite and Gurneyite wings of the Society. The scholar, Philip S. Benjamin, in his book, *The Philadelphia Quakers in the Industrial Age,* concurred with this judgment, calling the new thrust within the yearly meeting "the Haverford spirit" (p. 175).

Response to the Civil War

Neither Yearly Meeting made any reference to the American Civil War in 1861 or 1862, and in 1863 Race Street referred to the difficulty of bearing the peace testimony, while Arch Street remained silent. This did not mean, however, that Friends were not deeply involved in the war in a variety of ways. Some young Quakers, and some not so young, volunteered for the Union army, believing that the conflict provided an opportunity to end the evil institution of slavery once and for all. All members were forced to consider their faithfulness to the peace testimony, and to weigh that against their opposition to slavery.[10]

All three of the Friends publications contained much material about the struggle, both articles and editorials advocating a Christian peace witness, and factual information about the conduct of the war. Many Quakers did not subscribe to worldly newspapers and the weekly journals which came out of Philadelphia carried the only printed accounts they saw. Much of the writing on peace was a repetition of what had been said before, and Peter Brock tells us that

---

10. Peter Brock, *Pioneers of the Peaceable Kingdom* (Princeton, 1970), Ch. 6, "The Quakers in the Civil War." A paperback selection from Brock's larger work, *Pacifism in the United States: From the Colonial Era to the First World War* (Princeton, 1968). See also: Edward N. Wright, *Conscientious Objectors in the Civil War* (New York, 1961).

A Time of Change

"the war years were singularly lacking in any original contribution to the pacifist debate." Ezra Michener (1794–1887), a physician and amateur historian active in Race Street Yearly Meeting, published a pacifist tract in 1862 which was aimed at non-Quaker readers as well as his own co-religionists.[11]

As the war entered its third year the government decided it would need to resort to a draft, and Congress passed a Conscription Act in 1863 which made no provision for conscientious objection, although it did allow a person to hire a substitute or pay a $300 commutation fee. Most Friends were unwilling to adopt these alternatives, although there were individuals who cooperated. The pages of the *Friends Intelligencer* contained a lively controversy between a person signing himself N.R., who defended the practice of paying some commutation in lieu of service, and many other readers. He said that if the Discipline did not make provision for such commutations, it should be altered. Some subscribers felt the editor erred in printing his letters, and it is clear that neither *The Friend* nor the *Friends Review* would have been as tolerant. When a new conscription law in 1864 made provision for religious objectors to do alternative service in hospitals or with freedmen, or to pay a commutation fee which would be used to benefit the sick and wounded, this led to more difference of opinion. The Meeting for Sufferings of Arch Street refused to compromise in this manner. "Believing that liberty of conscience is the gift of the Creator to man, Friends have ever refused to purchase the free exercise of it, by the payment of any pecuniary or other commutation, to any human authority." While the editor of *The Friend* supported this uncompromising position, the *Friends Review* advocated accepting such alternative work provided by the government. Thus it is clear that Quakers were by no means united on how to interpret the peace testimony or on how to put it into practice.

In the meantime, Eliza P. Gurney (1800–1881), the widow of Joseph John Gurney who had returned to New Jersey after his death, felt led to revive the ancient custom of visiting heads of state in order to convey a religious message. Accompanied by three other

---

11. *A Brief Exposition of the Testimony of Peace, as exemplified by the Life and Precepts of Jesus Christ, and the Early Christians, and held by the Religious Society of Friends* (Phila., 1862).

Friends, she went to see President Lincoln in October 1862, to pray with him, and offer moral support. The Friends expressed the hope that slavery would be ended, and that peace would return as soon as possible. Lincoln responded warmly to this visit and kept up a correspondence with Eliza Gurney in the months which followed. He clearly recognized the dilemma which Friends faced, and promised to do all within his power to assist young pacifists whose problems were brought to his attention. According to tradition, Abraham Lincoln was carrying Eliza Gurney's last letter in his pocket the night he was shot at Ford's Theater. Quakers also went to see various members of Lincoln's cabinet, and some were far less understanding of the pacifist position, especially William H. Seward, Secretary of State. Edwin M. Stanton, the Secretary of War, who had some Quaker ties, had strongly supported the new provision for alternative service in 1864, but he was angered by the absolutist position of some Friends.

There are no hard figures on the number of Quakers who participated in the war, either by enlisting or allowing themselves to be drafted, nor are there accurate records of those who paid a commutation fee or had it paid by someone else. Students of the period agree that there were more Hicksites in the armed forces than Orthodox, but make it clear that a sizeable majority in both Yearly Meetings remained faithful to the peace position. There was a difference, however, in the way in which the two Yearly Meetings responded to the situation and it was reflected in the proceedings of the two business sessions in 1864. The Arch Street group referred to the conflict for the first time, as it discussed the sufferings of men who had been imprisoned for refusing to cooperate with the government. The closing minute, which was reprinted for general circulation, mentioned the suffering which existed in the country during this time, but did not mention the war specifically. In the State of the Meeting Minute at Race Street in 1864, reference was made to members who were serving in the armed forces, and Friends were asked not to engage in "harsh condemnation," but to pray that they would return home safely and that they would be led to accept the peace position of Friends.

Many Friends came under disciplinary action during the war for supporting the conflict in one way or another, and there were some disownments in each Yearly Meeting. However, the discipline was nowhere near as severe as it had been during the American Revolution. Virtually all Friends supported the government and goals of President Lincoln, and longed to see an end to slavery. In the earlier war, Quakers did not support the revolutionary government or its goal of independence from England. In addition, within the Hicksite body, a belief had evolved that each individual should be governed by his or her conscience, and not by some external authority.

Undoubtedly a number of the men who had gone off to the war were nominal Friends in the first place and never returned to their meetings. There were others, however, who felt strongly drawn to the old tradition, and such persons were willing to come to their meetings with an acknowledgment of wrong, and a promise to return to the testimonies of Friends. Race Street actively sought out such persons and called upon meetings to respond to returning veterans with a "spirit of restoring love." Some young men, in asking to be reinstated, pled extenuating circumstances for their behavior, and asked to be restored to membership without acknowledging wrongdoing. One man, who had been a brigadier general, claimed that the "inner light" led him to enlist, and he expected to remain a good Quaker despite his action. He wrote, "one of the most essential principles of Friends is obedience to conscience—much more essential than a belief in non-resistance."[12] Apparently Race Street Monthly Meeting did not challenge his position, for he remained a member until his death in 1909.

There was some debate among Friends about how far they should cooperate with the government in wartime. Some believed it was wrong to vote for candidates who were pledged to support the war, even questioned whether to support Lincoln in 1864. Once more the other issue of slavery carried a good bit of weight, and most decided to cast ballots. The matter of war taxes also came up for discussion. There was a tiny minority who refused to pay that proportion of their taxes which went to support the war, but most Friends continued to accept the principle of paying taxes "in the mixture," that is, in which various civilian expenses were covered in addition to military costs. An ultra-Conservative Ohio Friend, Joshua Maule, was the leading tax refuser of the war. Quakers found more unity in

---

12. Brock, *Pioneers,* pp. 293, 294.

opposing specific war taxes, which they classified along with paying a commutation fee to avoid conscription. The government felt no compunction about distraining goods to cover the cost of unpaid taxes, and no Friend went to court to challenge the right of the authorities to take such action.[13]

The war ended in early April 1865, when General Robert E. Lee surrendered at Appomattox. President Lincoln was assassinated a week later, just before Arch Street Friends convened for their annual meetings. The closing Minute, written on April 20, called it "a day of mourning and perplexity." When the Race Street body gathered in mid-May, it sent a memorial to President Andrew Johnson expressing sympathy on the death of Lincoln, and appreciation for the end of the war and of slavery. Two weeks later a small group of Hicksite Friends went to Washington to call upon President Johnson and members of the cabinet.

Aiding Freedmen and Others

Even while the war was still in progress, Friends in both Yearly Meetings began to launch various efforts to aid the freed slaves who were living in abject poverty in areas captured by the northern armies. Benjamin summed up their activities in these words: "Expression of the deep Quaker concern for black people which had been dammed up for over two decades by fears of war and violence came flooding forth as hostilities came to an end in 1864 and 1865" (p. 130). Women began to collect food and clothing, and later the men began to raise money. In the Race Street Yearly Meeting the men and women combined their efforts through the Association for the Aid and Elevation of the Freedmen. The Orthodox organization was known as the Friends' Freedmen Association.

The Orthodox group carried on its work in eastern Virginia and North Carolina, around Yorktown, Norfolk and other centers of what the army called "contrabands." The Hicksites concentrated their efforts in the Washington, D.C. region, and in eastern South Carolina. Friends in other yearly meetings, in the different branches

---

13. *Ibid.*, p. 321 ff.

A Time of Change

of the Society, undertook similar activities in the Carolinas, Tennessee, and Arkansas. British and Irish Friends gave generously to the work organized by Levi Coffin and those in the Gurneyite wing. Many other denominations rallied to meet this need, and the government created the Freedmen's Bureau in 1865. The initial task was to provide food and clothing for the men, women, and children who had been crowded into miserable camps by the army. Sometimes it was difficult to transport supplies to the camps, for the war had destroyed the railroads in much of the South, but it was possible to reach some areas by water.

Once the basic needs were being met to some degree, Quaker efforts turned to education. Southern states had made it illegal to teach slaves to read and write, for it was believed that education was likely to lead to dissatisfaction with servitude and, in turn, to insurrection or escape. The freedmen were eager for schools, and Friends set up simple classrooms as rapidly as teachers could be found. Dozens of persons went south in the early years to aid in the relief and education work, and a few stayed on for much longer periods. Cornelia Hancock (1840-1927) and Martha Schofield (1839-1916), from Race Street, taught in South Carolina for many years, and Yardley Warner (1815-1885), an Orthodox Friend, worked in Virginia, North Carolina, and Tennessee for a long period.[14]

By 1872 many of the schools had been turned over to government authorities, and the enthusiasm and generosity of Friends had begun to cool. The work of the Hicksites diminished more rapidly than that of the Orthodox, but the trend was the same in both associations. Other denominations lost interest even more rapidly than the Quakers. Some work continued in the following decades, and in 1887 Race Street appointed a new committee to undertake educational projects for blacks in the southern states. Anna T. Jeanes, who had been very generous in supporting the work of her Yearly Meeting, also created a trust to assist black education in the South. The Friends' Freedmen Association, in the other branch, continued on a reduced budget. It rallied to support the blacks who had settled in

---

14. See: Benjamin, *Philadelphia Quakers,* Ch. 6, "Tutoring the Freedmen;" and Rufus M. Jones, *The Later Periods of Quakerism* (London, 1921), Vol. II, "Friends' Work for Coloured Freedmen," p. 597 ff.

western Kansas just before the terrible droughts in 1879. They were forced to flee eastward, leaving all behind, when crops failed.

The members of each Yearly Meeting undertook a good deal of what was called philanthropic work during this half century, but the organization of such efforts differed between the two Yearly Meetings. Much of this effort was organized through separate associations by Gurneyite Friends in the Arch Street Yearly Meeting because the Wilburite Quakers were fearful lest activities planned by men and women might interfere with the working of the Spirit. While Race Street Friends shared some of this hesitation, it diminished as the years went by, and the Yearly Meeting was sponsoring a variety of good works by 1890. Members in each Yearly Meeting responded to the call from President Ulysses S. Grant for Quakers to aid in administering Indian reservations. Hicksite Friends responded through the Yearly Meeting committee on Indians, but the Orthodox joined other yearly meetings in creating the Associated Executive Committee on Indian Affairs. The Philadelphia Peace Association of Friends was organized by the Gurneyite wing in 1891, and Race Street Yearly Meeting formed a committee on peace and arbitration in 1897. Work for temperance among the membership was carried on in each Yearly Meeting, but Race Street created a special committee in 1891 which had a concern for public education and political activity. The Temperance Association of Friends was formed in 1881 by the activist members who centered their work at the Twelfth Street Meeting.

Friends in both branches had carried on various programs to aid the blacks of Philadelphia for many decades. While the government did very little in the nineteenth century, Benjamin concluded that as late as the 1860's the Quakers largely filled the needs of those unable to provide for themselves (p. 128). The plight of the newly emancipated slaves in the southern states captured the imagination of Friends for nearly a decade after 1863, but the needs closer to home drew support once more in the 1870's.

Friends met the needs of only a fraction of the poor in the postwar era, but they continued a variety of philanthropic works. Schools, programs for orphans, a home for older persons, sewing classes for women, a dispensary for those needing medical attention; these were typical of the things being done. These projects were all meeting needs, but they were not coming to grips with the basic

A Time of Change

social problems. The Institute for Colored Youth, which trained black teachers, and is the ancestor of Cheyney State College, came nearer to providing opportunities for fundamental change than most others. Supported by the Orthodox, this school was administered in a paternalistic manner which diminished the impact it might have had.

Quakers accepted segregation as a part of life, and, with a few exceptions, limited their philanthropic programs to all black or all white clients. They lacked appreciation for black culture, and tried to make the persons in their programs as much like whites as possible. These attitudes were shared by all middle class whites at the time, and Friends were no better, and no worse than others. In one regard they allowed their Quaker background to create additional friction. They refused to allow music in the schools they operated, including the Institute for Colored Youth, and they also excluded art. On the other hand, Friends made no effort to include blacks in their own meetings. They offered religious instruction both in Philadelphia, and in the South during the Reconstruction period, but not with the intention of making Quakers of blacks.[15] They were also excluded from Friends schools and colleges. In 1905 Swarthmore admitted a black student on the basis of his record without a personal interview, but when he appeared on the campus, he was politely rejected. Undoubtedly the same thing would have happened in the other institutions.

From Temperance to Total Abstinence

Initially, the concern for temperance continued to follow the pathway laid out in the previous chapter. In each Yearly Meeting the membership was reminded annually about the need for temperance, and in Arch Street a census of persons who deviated from the testimony was made each year. Members were not only urged to refrain from using alcoholic beverages, but to avoid any connection with the liquor trade, even to signing petitions of a person seeking a license to

---

15. See: Henry J. Cadbury, "Negro Membership in the Society of Friends," *Journal of Negro History,* 21 (1936), p. 151 ff.; Richard K. Taylor, *Friends and the Racial Crisis* (Wallingford, Pa., 1970).

Friends in the Delaware Valley

open a tavern. It was in this period that Friends moved from a testimony of moderation to total abstinence. Howard Brinton tells us that it was the younger men of Arch Street who persuaded the older leaders of the Yearly Meeting to adopt the more radical stand.[16] Fermented cider was added to the list of forbidden beverages during this same period. Race Street adopted the same position, but each Yearly Meeting continued to lament the fact that some members could not be persuaded to give up the old practices. Philip Benjamin found that initially rural Friends were more concerned about the problem than urban members, and that they associated it with their general fear and suspicion of the city. He also discovered that many of the leaders in the various groups had moved to Philadelphia from rural backgrounds. Nevertheless, as time went on, virtually all Quakers rallied to the temperance cause.[17]

Hicksite Friends combined efforts to persuade their own members to follow temperate practices with various programs to educate the public about the evils of liquor. In 1872 Representative Meeting memorialized the Pennsylvania government about the need for strict state control over the sale of alcohol, and in 1880 the Yearly Meeting wrote to Congress about the evils of the liquor traffic. In the 1880's conferences were held in each quarterly meeting to rally support for the temperance cause, and 6,000 copies of a pamphlet about the evils of cider were circulated. The committee on alcohol and tobacco cooperated with such organizations as the National Temperance Society and the Anti-Saloon League in carrying the message into public schools and in lobbying in Washington and in the state capitals. Friends in both Yearly Meetings supported the work of the Women's Christian Temperance Union in its early days, but withdrew from it later as it adopted more radical methods and causes in order to achieve its goals.

Arch Street Yearly Meeting seldom reached beyond its own membership in the temperance effort, although the Meeting for Sufferings sent an occasional memorial to state capitals about the evil traffic. In 1893 the Yearly Meeting gave up asking for a detailed accounting of every member who deviated from the total abstinence

---

16. Howard H. Brinton, *Friends for 300 Years* (New York, 1952), p. 143.
17. Benjamin, *Philadelphia Quakers,* pp. 94–97.

A Time of Change

position, but asked the subordinate meetings to continue a serious effort to persuade members to be faithful. In 1914 both Arch and Race Street sent memorials to Congress urging the adoption of the Eighteenth Amendment, the so-called Prohibition Amendment. The activist members of the Orthodox Yearly Meeting organized the Temperance Association of Friends in 1881, and through it carried on a number of projects in the wider community. The most vigorous temperance worker in the Yearly Meeting was Joshua L. Baily (1826–1916), a successful businessman in central Philadelphia, who threw both his time and money into the effort. He organized two temperance lunchrooms where workingmen could buy bread and coffee for five cents, as an alternative to the saloons which offered a free lunch with beer or other liquor. The Association cooperated with the same national bodies which the Hicksites supported, and Baily even worked with the Catholic Total Abstinence Union. The periodicals of both branches actively supported the temperance movement, although *The Friend* was more reticent about involvement with the world's people than the others. Rufus Jones wrote in the *American Friend*, "We honestly believe that the liquor problem is beyond all question the greatest problem now before the nation, and the greatest moral problem in the world."[18] This was the great crusade of the period. Racial issues had been pushed aside, and the great effort to work for peace had not begun. Economic issues had been intertwined with the liquor question in the minds of these reformers, and this question was agitated to a greater degree than ever before or since.

Support for the Doukhobors

In the 1890's Arch Street Friends learned from Friends in London about the persecution of a pacifist group in Russia called the Doukhobors. Tsar Nicholas II, angered by their refusal to perform military service, deprived Doukhobors of their homes in the Caucasus, and soon Tolstoy and other Russian liberals told the world about their plight. British Quakers, along with other groups interested in

---

18. *American Friend*, 5 (1898), Mar. 17, p. 244.

persecuted minorities, helped the Doukhobors to migrate from Russia, first to Cyprus and then to western Canada. Orthodox Friends, who were just beginning to reopen formal communication with British Quakers, were caught up in the effort to assist these people, and raised more than $30,000 in two years to pay for food, clothing, and shelter. The leading figure in the effort was Joseph S. Elkinton (1830-1905), a chemical manufacturer who also gave much time to religious work, and had been recorded as a minister.

The Doukhobors were not only pacifist, they followed no liturgical form in their worship, they did not observe the sacraments, nor did they follow the practice of employing a priesthood. To some sympathetic persons they seemed to be a bit like Russian Quakers. Actually, there were several ways in which they were very different from traditional Friends. They were communitarian in their economic practices, they pledged absolute obedience to their leader, Peter Verigen, and they incorporated much singing and chanting in their religious worship. In the beginning these differences were pushed aside as Friends concentrated upon providing for their physical needs, and convinced themselves that the Doukhobors, like the Quakers, depended upon a Christ Within for inspiration and direction.[19]

Elkinton and others persuaded the Yearly Meeting to accept responsibility for coordinating the efforts for the Doukhobors, the first time Arch Street was willing to undertake philanthropy for outsiders, aside from the traditional support of the Indian Committee. At first the Meeting for Sufferings provided oversight, but a special committee was named a few years later. Most of the money for the work was raised from Friends informally, although modest amounts were designated in the budget from time to time. Reports of the committee appeared in the Proceedings each year, along with those submitted from Westtown, the Indian Committee or the Book Committee.

The Doukhobors were hard working thrifty farmers and they were soon able to provide for their basic needs. When Friends turned from relief activity to education, they met some resistance, for the leadership was suspicious of schooling, fearing it would turn the community against the traditional ways of the movement. They

---

19. Benjamin, *Philadelphia Quakers,* pp. 115-119. See also: Joseph Elkinton, *The Doukhobors,* ... (Phila., 1903).

A Time of Change

feared, also, that the Quakers might attempt to inculcate Quaker beliefs and practices into the minds of the children. It became apparent that the Doukhobors were a stubborn, independent group who were determined to follow their own ways, and were unwilling to reach any compromise even with the Quakers. Some of the Doukhobors, those called "the Sons of Freedom," were radical anarchists who burned their buildings and marched naked across the land to protest government regulations. Friends found this behavior beyond their understanding, but slowly reached the conclusion that they were not going to be able to change them. In the years which followed, Quakers have gone to the assistance of the Doukhobors when their principles got them into trouble with the authorities.

A New Concern for Peace

If the work for the Doukhobors was done exclusively by the Orthodox, efforts to achieve peace in international relations and to uphold the peace testimony were shared by both branches. While the Disciplines of both Yearly Meetings were modified in a number of ways during this half century, the peace testimony was not weakened by either body. However, both Peter Brock and Philip Benjamin concluded that dedication to the testimony was not as great as it might have been, while agreeing that Arch Street was more diligent than Race.[20] Brock added that work for peace was often overshadowed by a concern for ". . . education, foreign missions and Sunday Schools, Indian work, temperance, and prison reform. . . ." *The Friend,* which paid less attention to some of those social concerns than the other papers, gave more space to peace than its competitors.

Such issues as militarization in the schools, the creation of compulsory state militias, minor military crises, and the need for arbitration of international affairs claimed the attention of each Yearly Meeting. Race Street created a committee on peace and arbitration in 1897 which had ties with the biennial conferences organized by Hicksite yearly meetings. While the Orthodox Yearly Meeting did not create a committee, it raised the issue in the annual sessions, and some-

---

20. Brock, *Pioneers,* pp. 340-348, 355-358; Benjamin, *Philadelphia Quakers,* pp. 192-195.

Friends in the Delaware Valley

times made direct reference to the need for personal commitment to the testimony in the State of Society summary at the conclusion of the Extracts. Meeting for Sufferings printed 60,000 copies of a tract entitled an *Address on War* in 1888. Race Street also circulated publications a few years later.

When the Orthodox Friends of the nation organized the Peace Association of Friends in America in 1867, Philadelphia did not participate because of its policy of isolation. Individual Friends cooperated with that organization and with non-Quaker ones in order to attempt to have a greater impact upon the public. In 1891 the Gurneyite group created the Philadelphia Peace Association of Friends in order to have a structure which would make the peace work more effective. These Friends met each spring during yearly meeting week, and carried on some educational and political activity during the intervening months.

Friends from both Yearly Meetings participated in the Lake Mohonk Conferences on International Arbitration organized by the New York Quaker Albert K. Smiley. In February, 1897, members of Arch and Race met together to support an arbitration treaty. Before his death, John Greenleaf Whittier (1807–1892) had told the British Friend William Jones that the Hicksites and Orthodox would someday get together to work for peace. He said that if they would work shoulder to shoulder, "they will learn to appreciate and love the good that is in all."[21] In 1901 the two Philadelphia Yearly Meetings hosted a peace conference for all American Friends. Professor William I. Hull (1868–1939) of Swarthmore, Isaac Sharpless and Rufus Jones of Haverford, Howard M. Jenkins (1842–1902), editor of the *Intelligencer,* were active in the peace movement at this time, along with businessmen like Charles Roberts (1846–1902), and Isaac H. Clothier (1837–1921). Some Friends were asked to support the peace testimony in other ways. Dr. Charles Huston, president of Lukens Iron and Steel Company at Coatesville, refused an order for "protective armor plate" as well as subsequent government orders, because it would have been inconsistent with his pacifist beliefs to support the military in this manner.[22]

---

21. Edwin B. Bronner, *The Other Branch: London Yearly Meeting and the Hicksites, 1827–1912* (London, 1975), p. 36.
22. Brook, *Pioneers,* p. 348.

A Time of Change

Friends had been outspoken in opposing the Spanish-American War, and they followed that effort with a defense of the right of the people of the Philippines to self-rule. They resisted the trend toward imperialism, and protested government actions frequently. Both Yearly Meetings wrote to President Woodrow Wilson in 1914 to protest the presence of American troops in Mexico. Friends cooperated in conducting a peace education effort in the non-Quaker colleges and normal schools of Pennsylvania. It is clear that Friends were in a much stronger position in regard to the peace testimony in 1914 that they had been before the Civil War.

Philanthropic Labor

Members of the two Yearly Meetings carried on various other philanthropic efforts during this period, sometimes through official channels, and often through some informal association. Both Yearly Meetings had a strong testimony against what was regarded as harmful literature. Quakers had always attempted to shield their children, and the adult membership as well, from both entertainment and printed material which might be harmful to the morals of the flock. Eighteenth century Quakers prevented theaters from operating in Philadelphia, and they had exercised censorship over printed materials. If Friends no longer had the power to control such matters, they could voice their support for efforts to suppress unfit publications. Race Street had a special committee on improper publications. It was only after World War I and the creation of the American Civil Liberties Union that Friends became strong opponents of censorship. The Hicksites also had a committee on Social Purity which was created to oppose prostitution. Both bodies were concerned about this social problem, for the center of the business in Philadelphia lay in the area between the Arch and Race Street meeting houses where there were supposed to be 300 brothels.

The First Day Schools which Friends started in areas where there were few if any members, sometimes blossomed into small settlement houses or other more modest efforts to improve the social and economic conditions of the persons who came to receive religious instruction. The Friends Neighborhood Guild is probably the best known of such projects. While the First Day School effort was under the direction of the Yearly Meeting at Race Street, the Gurneyite Friends carried on their work through the Friends First Day School Association. There were times when these two groups cooperated with one another in projects on both sides of the Delaware.

In another area of outreach, the mission effort in Japan, the Gurneyite Women Friends who began the work in 1883 received no support from the Wilburite wing of their Yearly Meeting, and expected no assistance from Race Street. Both the Wilburite and the Hicksite Friends were very much against missionary efforts of any sort, believing that it was undertaken in the will of human beings rather than the will of God.

In most respects, however, Friends had increased their awareness of social need, and began to make a contribution toward responding in places where assistance was desired. Benjamin concluded that even though Quakers ". . . were not in the vanguard of reform activity," they had made a good bit of progress by 1914.[23] If a half century earlier young Friends believed that a consistent Quaker was against many things, now the younger members felt that they were asked to be in favor of many things.

Friends Change and Move Toward Unity

The two Yearly Meetings undertook major revision of their Disciplines in the second half of the period under study, but characteristically, Race made the change in 1894, and Arch did not take similar action until 1910. Benjamin has emphasized the differences between the two compilations in his chapter entitled "Winds of Change," but one could just as easily stress the ways in which the Disciplines were similar in tone and content.[24]

Benjamin was correct in pointing out that disownment had been modified substantially by Race Street Friends, but it could be said that there was some easing of the practice in the other branch as well. The Hicksites included a new section entitled "Disownment," which

---

23. *Ibid.*, p. 190.
24. *Ibid.*, pp. 170–173.

A Time of Change

was really a plea to avoid disowning members if possible. "It is the desire of the Society that no one be disowned except when his retention would be to weaken our testimony for the Truth. . . ."[25] Arch Street Friends stated that the changes were "in tone more than substance," in their Discipline, but this was really true in both bodies. The new compilations included provision for disownment in each Yearly Meeting for violations of the peace testimony, and the testimonies against "oaths, trade, diversions, and the ministry."[26] There was a difference between the two documents about disownment for marrying out of meeting, but one gains the impression that the Orthodox, while maintaining this regulation, were beginning to overlook it, especially in some meetings. As early as 1891 the women of Twelfth Street Meeting requested a modification of this rule, but the Wilburite wing refused to allow the matter to be discussed during the revisions leading up to the 1910 edition.[27]

The Advices and Queries of the two Yearly Meetings were very similar in content and tone. The second query in each collection was identical, and the one about the education of children was very similar, although the numbering was different. The Orthodox had nine Queries, plus five additional ones, while the Hicksites had thirteen. Arch Street Friends included their Advices under the heading "Queries," while Race Street Quakers listed them separately as "Particular Advices." These statements, which were to be read on a regular basis in the local meetings reminded Friends of the need for right ordering in marriage, the peace testimony, temperance, simplicity, business dealings, care of members in need, receiving gifts and legacies, Bible study, and the holding of meetings for worship with a "free Gospel ministry." Neither set of Queries included a reminder of responsibility for one's neighbors, either near at hand or at a distance.

"If any one in membership with us blaspheme, or speak profanely of Almighty God, Christ Jesus, or the Holy Spirit, or shall deny the divinity of Christ, the immediate revelation of the Holy Spirit, or the inspiration of the Scriptures, he ought to be tenderly treated with for

---

25. *Rules of Discipline* (Phila., 1913), p. 105. I have used the 1913 edition of the Hicksite Discipline, and the 1912 printing of the Orthodox, the latest ones issued in this period.
26. Benjamin, *op.cit.*
27. *Ibid.*, p. 156.

his instruction, and the convincement of his understanding, that he may experience repentance and forgiveness."[28] This statement is from the Hicksite Discipline, not the Orthodox one. In 1827 the Orthodox had accused the other group of denying the divinity of Christ, a charge which had been stoutly denied at the time. During the intervening years since the 1860's, Race Street Yearly Meeting had stopped calling such denial a disownable offense, but it still declared its dedication to the statement. While the Orthodox used slightly different words and stated that rejection of this belief would lead to disciplinary action, the two Yearly Meetings were not as far apart as some believed.

A group of members in each yearly meeting founded a college during this period to provide higher education with a Quaker influence for the graduates of the secondary schools. The Hicksites obtained a charter in 1864 to create a coeducational institution, and Swarthmore College opened its doors five years later in 1869. The Gurneyite wing, rather than allow women to attend Haverford, worked together to found Bryn Mawr in 1885. An Orthodox Quaker architect named Addison Hutton (1834–1916), who designed Barclay Hall for Haverford, did the same for Parrish Hall at Swarthmore and Taylor Hall at Bryn Mawr. He was unusual in that he had personal friends in both branches of the Society.[29]

Elias Hicks had opposed higher education, and the Friends who were sympathetic with his views did not change their minds about the matter for several decades. When discussions about creating a college began, Lucretia Mott was one of the leading women Friends who urged that it be open to both sexes. Long a leader in many reform efforts, including women's rights, as well as clerk of the women's Yearly Meeting, Lucretia Mott had a strong influence among Hicksite Friends until her death in 1880. When Swarthmore opened, 85 percent of the students were in a college preparatory department, and it was some time before the new institution achieved recognition as a first rate college. Some members of the Board of Managers remained suspicious of intellectual excellence, and the administration was often caught between the desires of the

---

28. Compare the statements in the Race Street Discipline (1865), pp. 26, 27; (1913), p. 50; and the Arch Street one (1869), pp. 26, 35; (1912), p. 115.
29. Elizabeth Yarnall, *Addison Hutton, Quaker Architect* (Phila., 1974).

A Time of Change

students and those of the board. Dean Elizabeth Bond (1841-1926), who came to Swarthmore from Vassar, was a leading figure in changing the social and intellectual atmosphere on the campus. Jesse Holmes (1864-1942), brought strong reform concerns along with his rigorous intellectual interests to Swarthmore. Trained at Johns Hopkins in science, he did additional study in Europe before teaching philosophy at Swarthmore. In 1902 Joseph Swain (1857-1927) came to Swarthmore as president of the college. Formerly president of Indiana University, and earlier a mathematics professor at Stanford, he brought national recognition to the college during the next two decades.[30]

The men who organized the plans for Bryn Mawr were all closely tied to nearby Haverford, and they hoped to create a women's college comparable to the older institution where the young women of the yearly meeting might gain a greater appreciation of Quakerism while earning a college degree. The idea originated with Dr. Joseph W. Taylor (1810-1880) of Burlington, New Jersey, a medical doctor, business man, and concerned Friend who had accompanied the British Quaker J. Bevan Braithwaite on some of his journeys in the United States. Taylor left his estate to the project, and the main building was named in his honor. Dr. James E. Rhoads (1828-1895), another physician who was active in many reforms and editor of the *Friends Review,* was made president to carry out the goals of the trustees. They also named a Baltimore educator, M. Carey Thomas (1857-1935) as dean, and she had other ideas which she was able to put into practice.[31] Discriminated against by American universities because she was a woman, she set out to make Bryn Mawr a topflight institution for women, including graduate programs leading to the Ph.D. degree. Her Quakerism rested lightly on her shoulders, and she did little to maintain a Friendly influence in either the faculty or the student body. She became president in 1894 and made Bryn Mawr one of the top women's institutions in the country, but it ceased to have more than a faint, nodding acquaintance with Friends.

In 1914 both Swarthmore and Bryn Mawr enrolled some 450 students, while Haverford was less than half that size. All three were

---
30. Everett Hunt, *The Revolt of the College Intellectual* (New York, 1963), pp. 24-29.
31. Cornelia Meigs, *What Makes a College? A History of Bryn Mawr,* (New York, 1956).

recognized as important educational institutions, but the two with Gurneyite roots outranked the third at that time. Swarthmore had a very strong engineering program in those years, and was placing more emphasis upon athletics than some would have preferred.[32] The two yearly meetings were not sure that they approved the emphasis on the intellect at the colleges, fearing it might interfere with the Spirit. In 1903 Arch Street had warned Friends of the futility of attempting to enter the Kingdom through the intellect rather than the direct way pointed out by Christ. When George School opened, the Race Street Yearly Meeting expressed its belief that religious commitment was more important than intellectual achievement (1894).

During the early years of the twentieth century both Haverford and Swarthmore conducted summer schools where Friends could learn about some of the new ideas that were current, and consider how Quakers might respond to them. Higher Criticism of the Scriptures was considered, along with new scientific ideas and the concept of the Social Gospel. The history and beliefs of Quakers were also studied at these sessions. London Yearly Meeting had initiated such programs in 1895 with the Manchester Conference, followed by the Scarborough Summer School two years later. The impact of these educational efforts was augmented by participation in the biennial Friends General Conference gatherings for the Hicksites. In fact, the discussion of Christian Socialism at the F.G.C. sessions by such persons as Scott Nearing and William Walter Jackson upset some Friends very much. They did not realize that Jesse Holmes would be advocating the same thing in their own Yearly Meeting.

The Gurneyite wing of the Orthodox Yearly Meeting showed considerable interest in the other national grouping of Friends, the Five Years Meeting, but accepted the fact that Arch Street would not feel willing to join it. Four Quakers went to the Richmond Conference in 1887 which was the precursor of the Five Years Meeting, including James E. Rhoads of Bryn Mawr, and Henry Hartshorne (1823-1897), editor of the *Friends Review*. These men were welcomed as visitors, but did not share directly in the conference. Five years later Hartshorne and others went in the same way, and in 1897

---

32. Hunt, *op.cit.*, pp. 40-44; Benjamin, *Philadelphia Quakers,* pp. 43-45.

A Time of Change

a few individuals were invited, including Isaac Sharpless, but they were unable to attend.

In 1902, when the Five Years Meeting proper first convened, there were six Friends present from Philadelphia, plus Rufus M. Jones, whose membership was in New England Yearly Meeting. Some of these persons spoke, and a paper prepared by Edward M. Wistar about work with Indians was read to the gathering. This time three of the Philadelphians were women, including Elizabeth B. Jones, wife of Rufus Jones. The *American Friend* carried a report of the conference, but *The Friend* ignored it. The Wilburite wing was strongly opposed to the pastoral system, and some persons even had doubts about whether those persons were really Quakers. Five years later *The Friend* did publish an article about the Five Years Meeting, copying a report in the London *Friend* written by Gulielma Crosfield. At those sessions the Philadelphia participants were welcomed once more, and both Edward Wistar (1852-1941) and Isaac Sharpless were asked to address the sessions.

The Friend carried a letter in 1912, written by James Wood, chairman of the Arrangements Committee, inviting all interested Friends to come to the Five Years Meeting. A number of persons responded, and *The Friend* covered the session with an extensive article by William C. Allen (1857-1938). While he repeated earlier criticism of the pastoral system, he recognized the positive aspects of the Five Years Meeting which included all of the Gurneyite yearly meetings on the continent except for Ohio Yearly Meeting.[33] At the sessions in Indianapolis, Indiana, the gathering agreed to invite Philadelphia Yearly Meeting to join the others. Although the Yearly Meeting did not feel able to accept the invitation, it responded in a friendly fashion, and an informal relationship developed after 1914.

When Arch Street cut off all communication with other yearly meetings in the 1850's, it made one minor exception. For more than a century London and Philadelphia had maintained a policy of naming correspondents in one another's yearly meetings to oversee the exchange of travelling ministers, and this practice was continued even though they did not send epistles to one another. Generally

---

33. *The Friend,* 86(1912), Dec. 26, pp. 204-206.

speaking, the English visitors felt comfortable with the Gurneyite members of the Yearly Meeting, but they found it difficult to accept the quietistic attitudes of the Wilburite wing. Some of the visiting ministers were made very uncomfortable in the meeting which met at Arch Street as well as the one at Sixth and Noble. On rare occasions London sent an epistle to Philadelphia, hoping to reopen communication, but there was no response until 1897. That year Philadelphia did not send a specific epistle to London, but did send out a message "to all Meetings bearing the name of Friends and the members composing them." In 1906 London tried again, and this time had a reply from the clerk of Arch Street expressing appreciation for the epistle.[34] Soon a regular epistolary exchange was established after a break of more than fifty years.

At the same time Arch Street was responding to overtures from other Yearly Meetings, Race Street found London Yearly Meeting ready to open communications after seventy-five years of estrangement. Several things happened at once in the last years of the nineteenth century which made this change possible. London Yearly Meeting began to change from a predominantly evangelical body to a more liberal one which made it more tolerant of different approaches to Quakerism. Meanwhile, several of the Gurneyite yearly meetings which were officially recognized by London not only adopted the pastoral system of worship, but seemed to move toward a narrow fundamentalist position. When Joel and Hannah Bean, ministers of Iowa Yearly Meeting who were highly esteemed in Britain, were disowned by Iowa for holding unsound beliefs, many British Friends were indignant.[35] At the same time, many of the British Friends who travelled in America were favorably impressed by the Hicksites they met. The evangelical British-born minister Elizabeth Comstock, was only one of those who called for recognition of the Hicksites. She said, "I have attended many of their meetings, and am personally and very pleasantly acquainted with some of their leading members. I think that fully half of them are as orthodox as we are . . . We look into their libraries, and see all of our standard works. They bear as faithful a testimony as we do, to the spirituality of the New Dispen-

---

34. Bronner, *Other Branch*, pp. 56, 57.
35. *Ibid.*, pp. 43, 44.

sation, the freedom of the Gospel ministry, against slavery, war, oaths, conformity to the world, the sacraments, ordinances, and ceremonies."[36]

Members of Race Street Yearly Meeting who travelled in England at the turn of the century were welcomed at the Summer Schools, at Woodbrooke, and finally at Yearly Meeting sessions. Their authors began to appear in the *Friends Quarterly* and the weekly publications. The new biography of Elias Hicks, written in 1910 by Henry W. Wilbur (1851–1914) was reviewed in the journals. British Friends came to the realization that Hicksites were truly Quakers, and many began to think that they were preferable, in some ways, to some of the Gurneyites west of the Alleghenies.

During the early years of the twentieth century a few modest steps were taken toward healing the schism between the Hicksite and Orthodox yearly meetings.[37] The great bitterness which prevailed in an earlier period had faded by that time, and there were numerous examples of cooperation and friendship despite the official separation. The Friends Historical Association had been created in 1873 by four Hicksites and five Orthodox. Six years earlier Justus C. Strawbridge and Isaac H. Clothier had founded the department store which has been a landmark in Philadelphia for more than a century. Howard Brinton described the business partnership his father Edward Brinton and a prominent Hicksite named Herbert Worth formed in West Chester.[38] We have seen that Addison Hutton was an architect for both branches. When the Women's Medical College opened in 1856, Hicksite men and women were active in the oversight and management, and two of the early deans were Quaker women from Chester County, Ann Preston and Clara Marshall. By 1890 members of both branches were actively involved in the institution, as students and by providing support.[39] At the turn of the century the Hicksites developed the Buck Hill Falls Inn and residential community in the Poconos, and the Orthodox did the same, naming their

---

36. *Ibid.*, p. 35.
37. Thomas Speakman seemed to feel he should keep the old antagonism alive. See: *Divisions in the Society of Friends* (Phila., 1869, new edit., 1896), in which he called his branch "Friends," and referred to the others as "Orthodox."
38. H.H. Brinton, "Friends for 75 Years," pp. 3–11.
39. Benjamin, *Philadelphia Quakers,* pp. 153, 154; See: Gulielma F. Alsop, *History of The Women's Medical College . . .* (Phila., 1950).

community Pocono Manor. Some Friends apparently bought property in both developments, and a friendly athletic competition soon developed between the two vacation centers.

While Orthodox Friends were very slow to accept the idea of moving toward unity, Hicksite Friends were quite vocal in proposing such steps. The Race Street Yearly Meeting expressed approval of the idea as early as 1866. The *Friends Journal* often mentioned the idea and published material about the other branch, including western Quakers, and once it was joined with the *Friends Intelligencer,* the latter paper supported the idea. We have already seen that the first cooperative steps were in the area of peace work, but there was also joint effort in temperance. The young Friends were more enthusiastic about the subject than older ones, and both Horace Mather Lippincott and Henry J. Cadbury have written about the various conferences and gatherings which brought the younger Quakers together.[40] In local areas the two groups joined for historical celebrations. Organic unity was still four decades away, but the first steps had been taken, and the bitterness of an earlier period was fading.

Conclusion

In the half-century covered by this essay, the Quakers broke out of the immobility which had characterized the two Yearly Meetings since the schism of 1827. The body which convened at Race Street changed more rapidly than the one at Arch Street, but both Yearly Meetings had been greatly modified by 1914. While the quietistic majority was dominant in each group in 1861, the liberal element had taken over at Race Street by the turn of the century, and clearly had the initiative at Arch Street at the end of this period. Each group revised its Discipline, the Hicksites in 1894, and the Orthodox in 1910, which indicates the pace of change in the two bodies. More importantly, the attitude toward discipline and disownment also changed, and a desire to preserve traditions was replaced by an effort to find ways to respond to conditions in the new century. Friends

---

40. H.M. Lippincott, *The Struggle for a Union of Philadelphia Friends* (Phila., 1946), privately circulated. H.J. Cadbury, "The Development of the Young Friends' Movement in America," *Friends Intelligencer* 71(1914), June 13, pp. 369–371.

A Time of Change

were deeply involved with the world around them, they were participating in various reform organizations, and engaging in active political lobbying at all levels of government.

The Hicksite Yearly Meeting joined with other bodies in that branch of the Society to form the Friends General Conference. These gatherings introduced Quakers to many new ideas, it helped them to undertake new programs of social action, which they called "philanthropic labor," and brought spiritual guidance and support. After half a century of isolation, the Orthodox Yearly Meeting began to communicate with other groups of Friends, both the Gurneyites in the United States and Canada, and London Yearly Meeting as well as other bodies overseas. Because British Friends had also undergone a considerable change, they were able to put aside earlier prejudices against the Hicksites, and accept them as an integral part of the Quaker family. The change came slowly, first through personal visitation, and later in a more formal manner.

The Friends of Race Street founded coeducational Swarthmore College in the 1860's, and the Yearly Meeting opened George School in 1894. These two institutions made a profound impact upon Friends by introducing new intellectual and social issues which had not been considered before. Haverford College, with Gurneyite leanings, took a leading role in the other Yearly Meeting under the leadership of Isaac Sharpless, and Westtown School, despite its Wilburite flavor, also began to change. Bryn Mawr College was founded for women by the Gurneyite wing, but it soon moved away from any strong Quaker ties.

The bitterness and hostility which resulted from the schism of 1827 continued unabated into the early decades of this period. These feelings had such a negative influence upon the spirit of Friends in the two yearly meetings that one doubts whether any meaningful changes could have come in the two bodies without an easing of the antagonism. Individuals were the first to bridge the old hostilities, and the Hicksites were more willing to forget the past than the Orthodox, but a gradual thaw did take place. The two Yearly Meetings could cooperate on peace work more easily than anything else, and the creation of the American Friends Service Committee in 1917 early in the next period was a crucial step in this movement.

There were still many differences between the two Yearly Meetings in 1914, but a new spirit seemed to prevail, and this gave a new sense of mission and vitality to the Quakers of the Delaware Valley. In the Exercise of the Meeting at Race Street in 1908, one Friend was quoted as saying, "Unity does not necessarily mean agreement; indeed, it is not inconsistent with wide difference in opinion, expression and purpose. Unity is love, not likeness." Quakers were rediscovering what their ancestors had known, that love of one another is at the center of the Christian message. James Nayler had spoken of the Spirit which "delights to do no evil, nor to revenge any wrong, but delights to endure all things."[41] Friends were beginning to release that Spirit from the long imprisonment which resulted from the Separation.

---

41. London Yearly Meeting, *Christian Faith and Practice*, . . . (London, 1959), no. 25.

A Time of Change

Bibliographic Notes

Benjamin, Philip S. *The Philadelphia Quakers in the Industrial Age, 1865-1920.* (Phila., 1976).

Brinton, Howard. "Friends for 75 Years," *Bulletin* of Friends Historical Association (1960).

Brock, Peter. *Pioneers of the Peaceable Kingdom* (Princeton, 1970).

Bronner, Edwin B. *The Other Branch: London Yearly Meeting and the Hicksites, 1827-1912* (London, 1975).

Cadbury, Henry J. "The Development of the Young Friends' Movement in America," *Friends Intelligencer,* vol. 71 (1914).

_____, "Negro Membership in the Society of Friends," *Journal of Negro History,* vol. 21 (1936).

Diaries and Correspondence of Margaretta Walton . . . (1962).

Extracts from the Minutes of the Yearly Meeting Held in Philadelphia . . . (Phila., 1861-1914), (Orthodox).

Extracts from the Minutes of the Yearly Meeting Held in Philadelphia . . . (Phila., 1861-1914), (Hicksite).

Hunt, Everett. *The Revolt of the College Intellectual* (New York, 1963).

Meigs, Cornelia, *What Makes a College? A History of Bryn Mawr* (New York, 1956).

Rules of Discipline (Phila., 1912), Orthodox.

Rules of Discipline (Phila., 1913), Hicksite.

Sharpless, Isaac. *The Story of a Small College* (Phila., 1918).

Taylor, Francis R. "The Development of the Activities of the Yearly Meeting (1827-1927)," *The Friend,* vol. 101 (1927).

Wright, Edward N. *Conscientious Objectors in the Civil War* (New York, 1961).

Yarnall, Elizabeth. *Addison Hutton, Quaker Architect* (Phila., 1974).

IV Diminishing Separation: Philadelphia Yearly Meetings Reunite, 1915-1955

by HERBERT M. HADLEY

There was new life and vitality in the two yearly meetings of Friends in Philadelphia as they moved through the second decade of the twentieth century. Realization in both yearly meetings, Hicksite as well as Orthodox, that communication with London Yearly Meeting was being restored helped to put aside their sense of long isolation. Some younger members had experienced the companionship of English young Friends, both in England and at home. Most important, Friends of different ages had come to personal acquaintance with some Friends of the other branch. And such acquaintance had proved to be useful and enjoyable.

The Philadelphia *Evening Bulletin,* in early May 1916, commented on the recent action by the Yearly Meeting at Arch Street to prepare a letter of friendship addressed to the Yearly Meeting at Race Street—the first such letter ever written. In the words of the *Evening Bulletin,* "It is a remarkable fact that the schism which began when the 19th Century was still young should exist at a time when. . . all who participated in the original separation have passed into their graves." Noting that the differences between the two branches are not irreconcilable and could be bridged if there were general disposition on each side to do so, the *Bulletin* comment continues, "The moral influence which the Arch Street Friends and the Race Street Friends would have if they were to come together, would undoubtedly be much

Diminishing Separation

increased, and it is often a matter of wonder in the rest of this community that they have not acted on this consideration long ago."[1]

When Joseph Elkinton, a member and minister of Arch Street Yearly Meeting appeared at the Yearly Meeting at Race Street where that letter of good will was read, the (Hicksite) *Friends Intelligencer* described him as "an ambassador of peace." The opening words of Joseph Elkinton's personal message had been, "I am here because I love you."

The response by leaders among Race Street Friends was suitably warm. Joseph Elkinton's remarks were greeted "with outspoken approval" and Dr. Pusey Heald, of Wilmington, called Elkinton's visit "epochal." President Joseph Swain, of Swarthmore College, was prophetic in saying ". . . unity between all who bear the name of Friends will come out of the trauma of the great conflict now on in Europe."

One condition in both yearly meetings which would be changed in the coming years may be noted with this reference to Joseph Elkinton's visit to the Yearly Meeting at Race Street. Joseph had been accompanied by his wife, Sarah Elkinton, who was received by the women's Yearly Meeting in separate session. When Joseph went to the women's meeting, suitably accompanied by two members of the men's meeting at Race Street, the minutes of the women's meeting record that, "Coming as he did from a branch of the Society with which there has been no communication for nearly ninety years, Joseph Elkinton said that this visit, made in gospel love and under a real concern, required 'real heaven-born courage' to take the step he felt required. His message contained a desire for oneness in essentials—a right understanding of eternal realities. . . ." After prayer, the women's meeting records, the men left the meeting, and several women voiced their feelings of satisfaction with the visit paid and message received. Sarah Elkinton must have been silent until this time when she "was given kindly welcome, and fittingly responded."[2]

Separate yearly meeting sessions for women continued through 1923 at Arch Street and 1924 at Race Street.

A new enthusiasm and optimism led at least some Friends to think

---
1. *The Evening Bulletin* (Philadelphia), quoted in *Friends Intelligencer,* May 13, 1916.
2. *Friends Intelligencer,* May 20, 1916.

about their responsibility to reach out with a spiritual message to persons outside their meetings. Over several years, the Arch Street Yearly Meeting considered a concern for "those just outside our membership." *The Friend,* of January 2, 1915, carried a brief article which gave encouragement to "go out after the wandering ones (and) by the power of direct personal contact . . . win them to the service of the Master." In late March, the *Friends Intelligencer* reported, a series of meetings was held in Philadelphia by Isaac Wilson and others, and a member distributed, near the Fair Hill Meeting House in North Philadelphia, leaflets explaining non-credal and non-ritualistic religion. Later that year the *Intelligencer* carried a lead article with the title, "How to Increase the Membership of the Society of Friends." Its suggestions included: inform the public about times for Friends meetings and about what Friends are doing; make strangers welcome when they come to meetings, invite them to come again, and visit them; explain Friends' practices, especially the way of worship; have a good library; start clubs and associations which make use of the meeting house.[3]

The Foreign Missionary Association of Friends, constituted by members of Arch Street Yearly Meeting, was active in support of mission workers in Japan and China. In the spring of 1916 Dr. William Cadbury, whose life work was in China, gave an address to the Young Friends Movement, entitled "The Appeal and the Method of Foreign Missions." The editor of the *Friends Intelligencer,* Henry Ferris, wrote that Christian mission hospitals and schools do more to promote peace than any number of battleships, and "if we will now come out of our hermit-life, turn our faces to the sunrise, and join in the mission work which is fast making the peoples of the world one great family, we shall soon find millions waiting and longing to join our world-wide society of friends."[4]

Just one year earlier Joseph Platt, from a mission in Manchuria, wrote a letter to the *Friends Intelligencer* expressing disappointment that not one inquirer had responded to his call at the recent Friends General Conference for a teacher at the Union University in West China. He asked, "Are we holding up a conception of Jesus as He

---

3. *Ibid.,* May 22, 1915.
4. *Ibid.,* May 27, 1916.

wanted His disciples to hold? To say 'we have no creed' for some seems to say 'we have nothing.' What is the matter with us Hicksite Friends?" The letter from Joseph Platt brought response, from the editor of the *Intelligencer* and from its readers who were not ready to be pinned down to specific theological positions.

Nearer home than China or Japan, Friends of the two branches joined together to purchase and restore a house which had belonged to the family of John Woolman near Mt. Holly, New Jersey. Opened as the John Woolman Memorial House on June 10, 1916, this new achievement gave promise of strengthening Quaker life and witness, as the availability of historic Quaker sites in England had done for British Friends a decade earlier.

The Whittier Fellowship was an association in which young Friends of several yearly meetings participated. Among its leaders from the Philadelphia area were Henry J. Cadbury, Hannah Clothier Hull, Horace Mather Lippincott and George A. Walton. They held conferences in different meeting houses, reading papers written by their own members or holding discussions with invited speakers. At one time they expressed concern because earnest young people were seeking membership and wished to join Friends "in the root and not the branch." "Pilgrimages" were arranged, combining visits to outlying meetings with study, lectures, and social recreation. Summer Schools, at Haverford and Swarthmore Colleges, continued although some of these were now held off campus, as far away as New England.

The Woolman House School, which occupied a large house adjacent to the Swarthmore College campus, was opened in 1915 by the Advancement Committee of Friends General Conference. It was a training center for young persons who would take responsibility in their own meetings. Lectures were given by members of the Swarthmore College faculty and by Friends in nearby communities. After three years of service, Woolman House was handed over to a group of Friends from both branches, who would form the Board of Directors of Woolman School. The new Woolman School was to be a legally incorporated institution, with Elbert Russell as Director. It also became the forerunner of Pendle Hill.[5]

---

5. Carol R. Murphy, *The Roots of Pendle Hill,* Pendle Hill Pamphlet No. 223, 1979.

Friends in the Delaware Valley

In their local communities, where there were members of both branches of Friends, friendship and cooperation continued to develop. Celebrations of anniversaries, such as bicentennials of construction of meeting houses, were often the occasion for shared experiences. On one such occasion, Friends from both branches had come from near and far. The partitions dividing the house, and which were the only physical barrier between the local Hicksite and Orthodox Friends, were opened on this day. All sat together listening to papers and addresses from Friends of both branches. The spirit of love and unity was so strong that some Friends proposed that at the meeting for worship the next First Day the partitions should remain open, and Friends of both sides, could for at least one day, worship together. The proposal seemed acceptable, and it was thought to be adopted. But at almost the last moment word was received that one or two Friends were "uneasy" about the arrangement, and on their account it "seemed best" to close the shutters as usual.[6]

So it was with the two Philadelphia Yearly Meetings in 1916. While many were forward looking and ready to remove the dividing partitions, others were uneasy. It was necessary to wait—and to work together.

Cooperation in Peace Witness: The First World War

For the work which lay ahead of them the two yearly meetings had, or were soon to have, assistance from offices and staff. At Race Street, the office known as the Central Bureau had staff which served the yearly meeting and its committees: First Day School, Education, and Philanthropic Labor with its subcommittees. At Arch Street it was not until 1918 that the office of Secretary of the Representative Meeting was created, and in 1920 the title was changed to Secretary of Philadelphia Yearly Meeting of Friends. Management of the Book Store was included with the Secretary's organizational responsibilities.

In his book *The Philadelphia Quakers in the Industrial Age, 1865-1920,* Philip Benjamin says of Friends at the end of that

---

6. *Friends Intelligencer,* October 2, 1915.

period, ". . . in the terrible war [World War I] which followed, Friends seized upon their most precious testimony and, linking it to active service for the needy, charted a clear, concise course for themselves in the new century."[7]

Twenty years earlier, during the Spanish-American War, Friends had learned that their witness for peace was sadly diminished by their inability to speak with one voice on war and peace questions. Therefore, when their isolation from public affairs began to fade, concurrently with certain military actions and tendencies to military expansion by the United States goverment, the separate Friends bodies looked to other Friends groups who might join in a witness against military action.

There had been a Friends National Peace Conference at Winona Lake, Indiana, in mid-1915, which the *Friends Intelligencer* described as "one of the most important events since the separation of 1827." The principal resolution of the Conference called on the United States to convene a conference of neutral nations, pointing toward reform and extension of international law. It also addressed a message to all Christian Churches in America and an urgent letter to all meetings and members of the Society of Friends.

Friends first acted to assist financially British Friends who already in 1915 were caught in the calamity of war. To support the "relief work for non-combatants" Philadelphia Friends of both branches raised funds. There was still a strong desire on the part of many Americans, not only Friends, to stay out of the war which was raging in Europe. When "peace meetings" were arranged—for example in Burlington, New Jersey, by the Hicksite Quarterly Meeting's Philanthropic Committee—the principal speaker might be a Hicksite Friend while the chairman for the meeting was Orthodox. That meeting in Burlington had in its audience several businessmen and members of other denominations.[8]

The 1916 Yearly Meeting at Arch Street decided to address a letter to all monthly meetings of all branches of the Society of Friends, admitting in the letter that "for about seventy years" they had not corresponded with other yearly meetings of any branch. The Arch

---

7. Philip S. Benjamin, *The Philadelphia Quakers in the Industrial Age, 1865-1920* (Philadelphia, 1976), p. 212.
8. *Friends Intelligencer,* January 23, 1915.

Street Friends expressed "joy that Friends have already found so great a measure of unity in working for peace, their common heritage. We hope that unity in this great work may be further increased." They had appointed a permanent Committee on Peace, the Arch Street Yearly Meeting said, and would welcome—yes, and ask for—opportunity for cooperation with all similar committees.[9]

The Emergency Peace Committee of Race Street Friends, formed after the organization of the Peace Committee at Arch Street, at once joined enthusiastically in plans already mapped out. Whenever both branches of Friends were found in rural communities which the Peace Committees reached, there was "joyous, whole-souled working together." Thirty-nine meetings had been held in New Jersey, Chester County, Pennsylvania, Delaware, and Maryland. Audiences added up to about 5,500 persons. More than 20,000 pieces of literature were distributed.

The West Chester *Local News* of August 10, 1916, carried a story under the heading "Peace Automobiles." Automobiles filled with speakers on the peace issue, sent out by the two branches of the Society of Friends, arrived at the Unicorn Inn at Kennett Square. After listing names of the peace speakers and noting that William Wickersham had presided at the meeting, the news story said that other meetings would be held at Hamorton, Mendenhall, Fairville and Centerville.[10]

In November 1916, a "Joint Peace Meeting" was held at Fourth and Arch Streets with about fifteen hundred Orthodox and Hicksite Friends present. Rufus Jones and Leighton Richards, an English Congregational minister, spoke about the "tortures" which young men members of the Society of Friends in Britain were suffering for refusing military service.

Friends, again in 1916, addressed the United States government about its troops in Mexico, suggesting that the "police action" of keeping United States troops on the United States side of the border would be more desirable as prevention of crime than sending our troops across the border for punishing crime after it is committed.

The Philadelphia *North American* had in the summer of 1916,

---

9. *Minutes,* Philadelphia Yearly Meeting (Arch Street), 1916, pp. 14–15.
10. *Local News* (West Chester, PA.), quoted in *Friends Intelligencer,* August, 19, 1916.

offered its readers a ballot as a kind of referendum on which to indicate support of legislation requiring military training. The Peace Committees of the two yearly meetings gave wide distribution to instructions for interlining the ballot, making it read, "I am *not* an advocate of legislation for military training."[11]

But Japan must have seemed more distant, and the need for a peace message less urgent. From Gilbert Bowles, Quaker worker on behalf of Philadelphia Friends, had come a letter asking financial assistance for peace work in Japan. That letter had been presented by Race Street's Peace Committee at the Friends General Conference in June 1916, setting a goal of $600 as their share of that support. Six months later it was reported that only $285.79 had been received.

The Women's Yearly Meeting at Arch Street gave wide distribution to a Peace Message they had prepared in 1916. It deplored the trend toward militarism and war and declared, "There is a better way. . . work with us, not for a peace which means the mere cessation of war, the undisturbed opportunity to pursue wealth and ease—but for a peace charged with a positive, vital, energizing will to infuse into human society the Spirit of Jesus Christ."[12] The *Friends Intelligencer* reported that this Peace Message had been mailed to women's clubs all over the country, and that it had been printed in many newspapers including one claiming to have the largest circulation of any inland paper in Pennsylvania.

At the Arch Street Yearly Meeting in 1917 a statement signed by twenty-four young men and women was submitted, pointing to war and its connection with industrial and business life. The young Friends said, ". . . it is important that our Yearly Meeting should seek more fully to recognize what is implied by our stand against war. We need to discover what practical steps the Society of Friends should take in applying Christ's teaching of love and brotherhood in business, in the home, in politics and in all other relations of life."[13] This statement, and the devotion and sincerity of the young Friends who had prepared it, made a great impression on the Yearly Meeting.

---

11. *Friends Intelligencer,* August 12, 1916.
12. *Minutes,* Philadelphia Yearly Meeting (Arch Street), 1916, pp. 25, 29-33. *The Friend,* April 13, 1916, pp. 493-494.
13. *Minutes,* Philadelphia Yearly Meeting (Arch Street), 1917, pp. 13-15. *The Friend,* April 5, 1917, p. 481.

The way was thus prepared for new thinking by Friends about the social and economic order.

Meanwhile, as the United States seemed more certainly headed toward entering the war in Europe, Friends changed their attack on preparedness and militarism to an effort on preventing the country's direct involvement in the war. This was a less popular stand, and some Friends were not able to maintain their objection to this war. In late March 1918, the Philadelphia newspapers printed a letter headed "Some Particular Advices for Friends and a Statement of Loyalty for Others: Being the Views of Some Members of the Society of Friends Regarding Its Attitude Toward the Present Crisis." This letter, signed by 120 members of Hicksite meetings, in essence said, "*this* war is one in which Christians must give support to our national cause." One leader among Hicksite Friends, Charles F. Jenkins, had this to say: "Our members (who have signed this letter) desire to uphold our government in this awful crisis, but I do not want to change our peace testimony." A well-known Orthodox leader expressed what most Friends of that branch probably felt: "If the petition of 120 voices of the Hicksite meeting bears a representative tone of that branch of the Society, the Hicksites are surely facing a crisis in their history." Fortunately, the 120 voices did not express a majority, or any official, position within the Race Street Yearly Meeting.[14]

A "Friends Food Unit" was sponsored by a group of men from both yearly meetings. Horace Mather Lippincott described this project as a food production unit on land donated for the purpose by Walter H. Jenkins at Gwynedd, Pennsylvania. Fifteen Boy Scouts were engaged in planting and tilling in the effort to produce the greatest possible quantity of vegetables per acre, per worker. The work was supervised by an experienced gardener, and all was under the general direction of the W. Atlee Burpee Seed Company.

A "Peace Shop" was set up in central Philadelphia by women from the Hicksite branch. They were soon joined by Orthodox women, helping distribute posters and window displays as peace propaganda devices. A letter issued by women of both yearly meetings gave instructions for helping to provide materials needed for relief work and offered training courses in social services.

---

14. *Friends Intelligencer,* March 30 and April 6, 1918.

Diminishing Separation

On April 30, 1917, soon after the United States entered the European war, there was a meeting in Philadelphia of fifteen persons representing Friends across the United States. This group would coordinate the work of all Friends in America in connection with the "present crisis." An office would be set up in Philadelphia, and an executive secretary would be found. This was the beginning of the American Friends Service Committee whose story is told in another chapter.

Apparently New York Friends had first made the suggestion for a world conference of Friends in hopes of bringing peace. There had been some discussion about the advantages of holding the conference in Britain. By early 1917 British Friends had decided to invite representatives of all yearly meetings to take part in a conference in London, knowing that such a conference could not be held until the war was ended.

Friends in all branches were invited to correspond with an appointed conference committee, and to send agreed statements on subjects which should come before the conference. As the planning progressed six topics were selected for advance study by commissions, one set of commissions in Britain and another in America. That Philadelphia Friends played a major role in these studies is illustrated by listing the chairmen of the six study commissions in America:

  I. Fundamental Grounds of the Peace Testimony
     Rufus M. Jones
 II. Implications in National Life and International Relations
     Isaac Sharpless
III. Implications in Individual and Social Life
     William I. Hull
 IV. Problems of Education on the Peace Testimony
     Jesse H. Holmes
  V. Life of the Society of Friends in View of Present Demands
     Henry J. Cadbury
 VI. Methods of Propaganda
     Elbert Russell

Each of the six commission chairmen was a resident in the Philadelphia area as the conference preparations were made.

The first All Friends Conference was held in London, in August 1920. From Arch Street Yearly Meeting there were thirty-nine appointed participants, and forty from Race Street. Because there were still young men and women AFSC workers in different parts of Europe, some of these were able to cross from the Continent to England at conference time, adding to the number of official representatives.

For many members of both Philadelphia Yearly Meetings the conference was a "baptism" into the Quaker world. And as a capping climax to the tremendous peace effort carried on during the years of world war, this conference launched Philadelphia Friends, as it did others, into the struggle with evil in many kinds of social and economic problems in their own communities and around the globe.

New Awareness of the Social Order

Perhaps Friends in all their generations had been aware of the disadvantages of their poorer neighbors. Two reported instances of caring for the poor indicate attitudes and actions which may be commended. In the heat of summer 1916, the shaded grassy grounds behind the red brick walls at Fourth and Arch Streets were made available from 9:00 A.M. to nightfall for use by women and children from the surrounding slum streets. In that same summer Quaker "campers" at Pocono Lake took an interest in their mountain neighbors. Together they organized a Pocono Lake Improvement Association which included all the surrounding country. The County School Superintendent and the Farm Bureau Agent cooperated with the new Association, helping to start a rural center for general social uplift and to provide a visiting nurse for the area.

Wider fields of labor were indicated as needed in the statement, already mentioned, by twenty-four young men and women and presented to the Yearly Meeting at Arch Street in 1917. Those young Friends called attention to the link between war and certain kinds of practice in industrial and business life. The Yearly Meeting responded to this plea with concern to "discover how to interpret and put into practice [God's] will for the building up of a social and industrial order in harmony with the teaching of Christ." The young Friends who had presented the concern were made the nucleus of a committee to look further into the matter. Eventually, it was referred

Diminishing Separation

to a committee appointed a year earlier, known as the Social Order Committee.

One year later (1918), a small pamphlet, "A Message from the Social Order Committee of Philadelphia Yearly Meeting" (Arch Street), attempted to provide a definition of "social order." The social order, by this definition, embraces "all business relations, including those between buyer and seller, employer and employee, borrower and lender, owner and renter, and the relations of each of these to the community and to the state. It also includes relations between what are commonly called classes of society."[15]

A very comprehensive definition of the social order! The committee's pamphlet proceeded to enumerate ideals for which Friends should strive and suggested initial steps toward realization of those ideals. One of the suggested steps was "the making of investments in the spirit of service rather than of self-interest, investigating, as far as possible, the industrial conditions lying back of securities and favoring those investments that have a social motive, even if returning a low rate of interest."

Friends in both yearly meetings were enthusiastic about the possibility that a new social order was arriving. William Eves III, wrote about meetings of the two branches in Wilmington joining together for mid-week discussion of important topics with able leaders. The first topic was "The Need for the Study of the Social Order at the Present Time." Late in 1918 the Race Street Yearly Meeting's Committee on First Day Schools used a section of the *Friends Intelligencer* to give instructions about organizing classes and subjects for study. In the December 28 issue the study of the new social order was introduced by this paragraph, "One reads scarcely a newspaper, periodical, or presidential message these days without noticing the frequent occurrence of such expressions as 'the new social order,' 'the passing of the old regime,' 'the rising tide of democracy,' 'the end of autocratic power,' 'control through self-determination.' The reader knows full well that we are standing at the threshold of a new era in civilization. . ."

At least some Friends seem to have held the belief that a new social order would actually arrive after the world war was ended, because it

---

15. *Minutes,* Philadelphia Yearly Meeting (Arch Street), 1918, p. 20.

would be so obvious that wrong social order had brought on war and that people and institutions would be ready to change. The *Friends Intelligencer*'s editor, Henry Ferris, always optimistic, wrote on December 14, 1918, "If the ideas of Jesus are to be the basis of the new social order, the kings of wealth will go the way of the kings of power. No matter what the nominal form of the government may be, in the new social order the leaders will be men of a new type." Grasping at a bit of realism he added the question, "Are we ready for it, though?"

In his book Philip Benjamin points to certain Friends who made contributions in specific areas of the social order. Morris E. Leeds proposed a modified profit sharing plan for industry. Agnes L. Tierney urged Friends to make common cause with workers and to examine the injustices of contemporary economic conditions. She also urged investment of inherited wealth in worthy enterprises such as housing projects and "garden cities," in spite of low return. Henry Albertson recommended that Friends abstain from the purchase of land for speculative purposes, and that affluent Friends should not leave money to the young who would thereby be given unfair advantage over less fortunate youth. Morris Leeds and M. Albert Linton were influential in arranging sponsored meetings to explore ethical questions relevant to social, economic, and political problems with such persons as Jane Addams, Frederick Howe (a municipal reformer and immigration commissioner), and Glenn Plumb (an advocate of railroad consolidation).[16]

These and other Philadelphia Friends took a very active part in the development of the American Friends Service Committee's role in applying Quaker principles to the social order. The AFSC's developing Social and Industrial Section was guided by Bernard G. Waring, its chairman, who was also chairman for many years of the Social Order Committee of the Arch Street Yearly Meeting.

Closely related to questions of the social order were new concerns for: civil liberties, growing out of Friends' work during the war with conscientious objectors; a greater sensitivity to their testimony for equality; new manifestations of simplicity in life styles rather than in outward form such as the plain dress; and changed ideas of "com-

---

16. Benjamin, *Philadelphia Quakers,* pp. 209–210.

Diminishing Separation

munity" which impelled Friends into the world rather than protecting them from it.

Women had gained a status in Quaker organizational structure which was far more equitable than that to which they had been confined in separate but unequal Friends business meetings. In 1918 the Race Street Yearly Meeting officially gave support to the proposed constitutional amendment granting full suffrage to women and urged the senators from California, Delaware, Maryland, New Jersey, and Pennsylvania to vote for the amendment. (Why California? The Orange Grove Monthly Meeting in Pasadena was at that time a part of the Philadelphia Yearly Meeting, Race Street.)

From the Social Order Committee there developed a Women's Problems Group which later became an independent body with its own program, continuing into the 1960's.

The separate yearly meeting sessions for women came to an end in the 1920's. At Arch Street in 1923 the men and women met in joint session on two days and in separate session on three days. At the beginning of Yearly Meeting in 1924 the decision was made that all sessions that year would be held jointly. Race Street was not far behind in this change. In 1923 the women's meeting had joined with the men for one session only, while all other sessions were separate. One year later only one session was held separately and in all others men and women met together. For several more years the minutes of both Yearly Meetings show that they met in "joint session"; i.e., women and men together, until finally participants of both sexes were considered one Yearly Meeting at both Arch Street and Race Street.

Arch Street's Social Order Committee had been formed in 1917 and its work was directed toward change in the social order. The long-standing Race Street Committee on Philanthropic Labor, with its many facets of philanthropic activity designed to minister to the needs of disadvantaged persons, found it difficult to accommodate the new concern for a changing social and economic order.

During the First World War Race Street had appointed a Committee on Peace and Emergency Service which, in 1918, added to its structure a Social Order Department which was instructed by the Yearly Meeting "to work as way opens with similar organizations of the Arch Street Friends or other organizations of Friends."

Friends in the Delaware Valley

Again in 1927, the Race Street Yearly Meeting proposed to form a Social Order Committee for cooperation with the committees of London Yearly Meeting and the Yearly Meeting at Fourth and Arch Streets. The proposal was referred to the Committee on Philanthropic Labor to proceed as way opens and to report the following year. The difficulty which confronted the Race Street Friends involved money. In 1935 the Philanthropic Labor Committee was once more asked to prepare a proposal which might require a change of name to "The Committee on Philanthropic Labor and Social Order" and to refer to it briefly as the Social Order Committee. In this way it was thought the legal standing of the committee would be preserved to entitle it to income under the will of Anna T. Jeanes, yet permit closer cooperation with the Arch Street Committee in name as well as in work.

In that year (1935) this minute was recorded at Arch Street: "We accede to the request that our Social Order Commmittee be allowed to unite with a similar committee likely to be appointed by Race Street Yearly Meeting to form a Joint Social Order Committee of Philadelphia Friends."

The philanthropic idea or terminology at Race Street was tenacious. During a further decade there was a struggle over the concept of "social service," which Race Street wanted in the committee's name, and "social order," to which Arch Street Friends were committed. Finally in 1948, the Arch Street Yearly Meeting heard a report from its Social Order Committee which included this: "We are glad to hear that the Race Street Committee on Economic Problems has changed its name this past week to the Social Order Committee to indicate their practical identity of purpose with our Committee."

In 1941 there was recorded in Minutes of Arch Street Yearly Meeting a report of the Young Friends Movement which said that value had been found in a new and important project sponsored by the Social Order Committee—the weekend workcamp at the Wharton Settlement in Philadelphia. The report continued ". . . physical work, study and meditation together have deepened the spiritual experience and increased the sense of community among those who have participated."

A brief account of each of three concerns, relevant to the social order and particularly relevant to the times in which each concern

Diminishing Separation

arose may be given here. In 1920 and 1921 Arch Street Friends were especially disturbed by the kinds of amusements which attracted young people, including their own members, as isolation from worldly affairs decreased. An appointed committee arranged a conference, which was held in May 1920, of parents, teachers, overseers, and others interested in the training of young people. Such amusements as movies, dancing, and theater-going were seen as incompatible with a progressive, deepening spiritual life. "Should not we, who have the calling of Friends, reach beyond the limitations of the unthoughtful and make our recreations and social life positively strengthen and foster the whole purpose of our lives?"[17]

In 1927 a controversy arose over the dismissal of a Friend, Robert T. Kerlin, from his position as Head of the English Department in the State Normal School at West Chester. Four years earlier the *Friends Intelligencer* had printed a long, and very able, exposition of Friends' principles which had accompanied the application for membership in Friends meeting by the same Robert Kerlin whose poetry also frequently appeared in the *Intelligencer*.[18] But the American Legion had publicly accused Robert Kerlin, a pacifist, of (1) criticism of certain foreign policies of the administration now in power in Washington; (2) friendly social relations with Negro teachers; (3) opposition to war, and "radical opinions." That some number of Friends agreed with the American Legion, in greater or lesser degree, is obvious from letters to the editor of the *Friends Intelligencer*. Others supported Robert Kerlin, and fortunately his Concord Quarterly Meeting came strongly to his defense. Concord's minute was forwarded to Race Street Yearly Meeting which "deplore[d] the use of violent denunciation, social ostracism, and harsh interpretation and execution of law in an endeavor to control the thoughts and ideas of others." The Yearly Meeting asked its Representative Committee to have "care of this situation in general and of individuals under attack" and to act as way opens.[19]

Similar witness was made by the two yearly meetings jointly in 1954, the "Joe McCarthy era," when they adopted "A Statement on Civil Liberties" which said, in part, "[We are] disturbed at increasing

---

17. *Minutes,* Philadelphia Yearly Meeting (Arch Street), 1921, pp. 17–20.
18. *Friends Intelligencer,* January 6 and 13, 1923.
19. *Ibid.,* April 16 and 30, 1927. *Minutes,* PYM (Race Street), 1927, p. 6.

encroachments on the freedom and integrity of the individual by irresponsible accusations, by pressures for conformity in thinking, by charges of guilt by association, insistence on assertions of loyalty, and by the assumption of guilt rather than the presumption of innocence. We reaffirm our belief in man's integrity; we must reawaken in our fellowmen a real faith in their spiritual nature."[20]

In the early 1930's, the years of the Great Depression, the Arch Street Social Order Committee began to call attention to the problem of unemployment. The first efforts were to reduce the numbers of unemployed. In 1934 a Joint Commission of the two yearly meetings was charged with responsibility to study the problems and possibilities of Unemployment Insurance. By yearly meeting time in 1935 they had completed a pamphlet entitled "Unemployment Compensation Plan" and work was progressing on kindred subjects.

At different times the Social Order Committee had subcommittees on Rural Problems and on Friends Social Work. For many years it sponsored a Business Problems Group. In 1950 it presented at a joint session of the two Yearly Meetings a statement on "Peace and Social Justice" which was approved by each Yearly Meeting in separate sessions.

Uniting for Peace—A Joint Committee, 1933

The enormous work for peace which was done by Friends of both yearly meetings during, and immediately after, the First World War has already been described. Arch Street had named a permanent Peace Committee and Race Street had made appointments to fill the "war emergency" period. "Without compromising the identity of the two committees, much helpful cooperation has been enjoyed in holding meetings, publishing and distributing literature and gaining proper publicity on important questions," said the report of Arch Street's Peace Committee in 1917.

The representative bodies of the two Yearly Meetings authorized a Quaker delegation to call on President Harding at the White House. The delegation, including seven Arch Street and two Race Street

---

20. *Minutes,* PYM (Arch Street), 1954, p. 63.

Diminishing Separation

Friends, saw the President on March 23, 1921. Their principal concern was for disarmament. They found Mr. Harding firmly committed to a strong navy, but reported it seemed reasonable to hope that he would throw his influence toward some international agreement. A few days after that visit to the White House, the Arch Street Yearly Meeting authorized a letter to the President asking him to convene in the near future an international conference on disarmament.[21]

A member of the Race Street Peace Committee put forward in 1923 a concern that a letter be addressed to "Fellow Christians and Sister Churches in All Lands." Such a letter was prepared by the Arch and Race Committees acting together, and was endorsed by London Yearly Meeting and by several other American yearly meetings. It was distributed widely throughout the world in the spring of 1923.[22]

Arch Street Yearly Meeting Minutes in 1933 record approval of a plan under discussion to form one Peace Committee of Friends in the Philadelphia area. This was accomplished in the spring of that year. The Peace Committee's office was located at 304 Arch Street; its lending library and information center was set up at 1515 Cherry Street.

By 1935 the two yearly meetings were holding their annual sessions in the same week of spring. That year in separate sessions on the same day they authorized a public meeting to be held in their name at Reyburn Plaza in Philadelphia to present Friendly testimony that disputes can be settled by peaceful methods rather than by armaments and rousing of a warlike spirit. This action was taken to counteract a display of war equipment and war maneuvers that same day at Reyburn Plaza.[23]

A concern for exemption from payment of war taxes was brought to the Race Street Yearly Meeting in 1941 by Margaret E. Dungan, a member of Providence Monthly Meeting. She hoped that way might be found to enable members of religious bodies conscientiously opposed to war to be exempt from payment of taxes earmarked for military use, provided an equal sum was paid by such persons into a government fund for humanitarian purposes. Friends were sympa-

---

21. *The Friend,* March 13, 1921. *Friends Intelligencer,* April 2, 1921, p. 213.
22. *Minutes,* PYM (Arch Street), 1923, pp. 23, 78.
23. *Minutes,* PYM (Race Street), 1935, pp. 13-14.

thetic to this concern, but "until the full implications of the proposal are studied formal action by the meeting would be premature" said the Race Street minute. This concern, and others related to war and peace as they arise, would be referred to the Joint Peace Committee (if Arch Street Yearly Meeting were to concur) for thorough study and report at a subsequent session of Yearly Meeting.[24]

Forming Other Joint Committees

A variety of concerns began to receive more organized attention in the years immediately after the First World War, and many of these came into the care of a committee.

Missions—Emphasis on Japan. A number of Arch Street Friends had keen interest in Christian missionary efforts. In 1918 that Yearly Meeting appointed a committee "to consider our responsibility to work in foreign lands." Japan received special mention, but interest was shown also in China, India, Syria, and the Sudan. This new committee was not expected to administer any missionary project, but it would try to pull together the mission interests of Friends in several independent agencies. In 1923 Arch Street authorized the setting up of a Missionary Board, combining the interests of their official Committee on Christian Labor in Foreign Lands and the independent Foreign Missionary Association of Friends, the latter being responsible for the Friends' work in Japan.

By 1947 the work in Japan had become the only overseas undertaking of Philadelphia Friends, and its nature had changed to cooperation with an indigenous yearly meeting of Japanese Friends and support for the Friends Girls School in Tokyo. In that year an invitation was given Race Street Yearly Meeting to join fully in this work through a joint Japan Committee. This was achieved in 1951.

Temperance. A testimony for total abstinence from use of alcoholic beverages had developed through the nineteenth century. In both of the Philadelphia Yearly Meetings there was an annual accounting of the number of members who failed to keep to this standard. Replies to that query, as provided by the monthly meetings, indicated that most members did abstain. The few who did not

---

24. *Ibid.*, 1941, p. 11.

Diminishing Separation

were, apparently, moderate drinkers. As the opposition to manufacture and sale of alcoholic beverages increased across the nation, Friends actively supported the effort to add a prohibition amendment to the United States Constitution. This was achieved in 1919, and the following year the Arch Street Yearly Meeting discontinued use of the query which called for an account of members' use of alcohol. The subject was committed to the care of overseers.

With repeal of the prohibition amendment in 1933, the yearly meetings again became more actively involved in work for temperance. The long-standing, independent Friends Temperance Association had requested and received appointments to its executive committee by both yearly meetings in 1932. In effect, the Temperance Association had become a joint committee of Philadelphia Friends. It promoted temperance education among Friends and other groups. In 1943, with social conditions disturbed by World War II, the Arch Street Yearly Meeting "united with Race Street Friends in recognizing the need for continued and greatly increased" effort in this concern. The Representative bodies were asked to study the situation, searching for helpful ways of advancing the temperance cause. The Young Friends Movement was asked to select six of its members to assist with this effort.

Race Relations. The very considerable involvement of Friends in work with blacks freed from slavery at the time of the Civil War had diminished to some support by Hicksite Friends for a few schools for blacks in southern states. In Philadelphia there was philanthropic service to poor blacks as well as to poor whites, but little recognition of need for social change to remove injustices suffered by Negroes.

A concern was brought to the Arch Street Yearly Meeting in 1919 about lynchings in parts of the South, the number of lynchings having noticeably increased. The Yearly Meeting responded by appointing a Committee to Protest Against Lynching. It was continued during a second year. Some of its members travelled into the South, arranging appointments with state governors and law enforcement officers to urge greater diligence in application of laws which should protect all citizens from physical violence. The committee reported that their members had been generally well received when undertaking such visits to officials, and in 1921 it was agreed this committee should be laid down.

Its most active member, Esther Morton Smith, spoke to the

Friends in the Delaware Valley

Women's Yearly Meeting that year about her wider concern for better relationships "between white and colored peoples." She spoke of bitterness of feeling among many blacks toward white people because they suffer injustice and deep prejudice from many whites. The situation called for efforts by Friends "to substitute Christ's spirit of love and fellowship and mutual helpfulness, for that of unreasoning hatred and prejudice." Esther Morton Smith was persuasive. The Arch Street Yearly Meeting appointed a Race Relations Committee that year (1921), on which she served as chairman for many years.

The activities undertaken, and the particular incidents reported by Arch Street's Race Relations Committee seem barely to reach the threshold of actions required to bring anything approaching equality of the races, although Chairman Esther Morton Smith had a very open and understanding mind and spirit. At the same time, the Race Street Committee on the Interests of the Colored Race began to recognize that changing times and developing concern called for change of its name. That was done in 1929; the new name was simply Race Relations Committee and it began immediately to function jointly with the Arch Street Committee. The Joint Committee's first annual report, in 1930, noted that Philadelphia had the second largest "colored population" of any city in the United States, and that this fact gave special significance to the committee's work.

The joint Race Relations Committee in 1932 made reference to the particular problem of increasing segregation, and reported on the committee's active educational program in regard to this question. The committee was concerned "to arouse the Society of Friends as a whole to a sense of their greater responsibility in the race situation."

Friends were very soon put to the test, and were found unprepared. Along with consideration of the joint Race Relations Committee report at Arch Street in 1934 a statement from the Westtown School Committee was read. Application for admission to Westtown School had been received from two promising Negro boys, but there was not unity about this matter in the School Committee, and the two boys were denied admission. However, the Scholarship and Admissions Committee had been "instructed to study the whole problem of the admission of a small, highly selected group of pupils representing other races and cultures."

The Westtown Committee told the Yearly Meeting that the problem was larger than the admission of Negro students to the School. The attitude on race relations of the entire yearly meeting was relevant. The issue would ultimately have to be settled by the Yearly Meeting as a corporate body. The first black student at Westtown was admitted in 1945. At Race Street Yearly Meeting's George School this step was taken in 1946.

Prisons and Prisoners. A desire that the Representative Meeting should give "earnest attention" to the jails in Pennsylvania, New Jersey, Maryland, and Delaware was expressed in Women's Yearly Meeting at Arch Street in 1920. In 1936 Race and Arch Street Yearly Meetings collaborated in preparation of a statement reaffirming their belief in the sanctity of human life, and urging legislation to abolish capital punishment. In 1954 the Women's Problems Group presented to a joint session of the two yearly meetings their concern about prisons and prisoners, urging the appointment of a joint yearly meeting Committee on Prisons. This step was approved.

Marriage and Divorce—Family Relations. The question of divorce came before the Arch Street Yearly Meeting in 1919. The matter was referred to Representative Meeting with concern "to quicken public sentiment against the divorce evil. . . ." A year later the Yearly Meeting decided not to say anything about legislation on divorce since, from a religious viewpoint, this question was one to be dealt with by education and quickening of the spiritual life. There was some feeling that the Discipline's advices on marriage could be advantageously rewritten "so as to appeal more practically" to younger Friends.

By 1932 the Arch Street Representative Meeting had prepared a "Statement of Ideals and Counsel on Marriage and Divorce" which encouraged emphasis on positive, loving relationships in marriage and blamed the increasing incidence of divorce—even among Friends—on the prevalent claim to individual happiness as a right and the unwillingness to sacrifice some of one's own interests to the interest and welfare of others. The Yearly Meeting appointed a committee of fifteen persons to study the implications of the divorce problem and to present the results of their study to the meeting a year later. The result was the appointment in 1933 of "a board of professional and lay persons" as the Marriage Council of the Society of Friends whose functions would be education, personal advice, and collection of factual data.

The Marriage Council must have functioned well, for in 1944 its report in Race Street Yearly Meeting Minutes said it had in effect been a joint committee of both meetings for several years through the appointment of members of the Race Street Social Service Committee to work with it. Since Race Street Yearly Meeting in 1943 had authorized the appointment of fifteen members to the "Marriage Council of the Philadelphia Yearly Meetings," it was in fact a joint committee.

Work with Indians. The Race Street Yearly Meeting in 1948 accepted the invitation, issued two years earlier, of the Indian Committee at Arch Street to appoint members to a joint committee. The Race Street Friends already appointed representatives to the continent-wide Associated Executive Committee of Friends on Indian Affairs. These representatives were made their members of a Joint Indian Committee of the two Philadelphia Yearly Meetings. An account of the work with Indians is provided in another chapter of this book.

Ministry and Worship. The Executive Committee of Ministry and Counsel (Race Street) made its first report to Yearly Meeting in 1934. It reported that a growing spirit of fellowship and unity had developed between their Executive Committee and the Arch Street group of Ministers and Elders. This fellowship had been strengthened by the assistance of concerned members of the Young Friends Movement. And in 1936 the Arch Street Yearly Meeting appointed a Committee of Ministers and Elders to Cooperate with Race Street Friends.

Religious Education. As early as 1918 there was held at Friends Select School a Friends Conference on Religious Education, arranged by a committee representing the Education and First Day School interests of both yearly meetings. Similar arrangements for conferences were made in subsequent years. While Race Street Yearly Meeting and the Friends General Conference had active religious education programs beginning in the nineteenth century, Arch Street Yearly Meeting did not appoint a Religious Education Committee until 1929. That committee's first annual report recalled that some members of that Yearly Meeting—they were of the Gurneyite wing—had started the Friends First Day School Association sixty-nine years earlier, with principal concern to reach non-members. Long since, however, there had been classes for members in their own meetings. The Religious Education Committees of both yearly meetings joined with the Young Friends Movement to sponsor the William Penn Lectures.

Education and Friends Schools. Each yearly meeting had its committee for management of a boarding school—Arch Street for Westtown and Race Street for George School. For assistance to the other Friends schools each yearly meeting had an Education Committee which fostered, respectively, the Friends Educational Association (Arch Street) and the Association of Friends Schools (Race Street). At Race Street Yearly Meeting in 1927 there was a report that "joining with Arch Street Friends in the management of certain Friends schools has resulted in increased attendance."

A Conference on Education held at Germantown in 1928 recommended that the Yearly Meeting at Arch Street instruct its Educational Committee to cooperate as way may open with Fifteenth and Race Street Friends, suggesting that, its scope thus enlarged, the committee be renamed "Council on Education." This was approved in general, and Arch Street members were named. The proposal to Race Street was approved there in 1930, and eventually in 1933 there was a Friends Council on Education with representatives appointed by both yearly meetings. The yearly meetings' Education Committees remained separate but cooperating units; it was reported in 1949 that their libraries had been merged.

The Important Role of Young Friends

Through all of the activities which brought members of the two Philadelphia Yearly Meetings into personal association, then into informal cooperation and on to joint undertakings of an official nature, young Friends had a very considerable part. The *Friends Intelligencer,* commenting editorially on the Yearly Meeting held at Race Street in 1915, said that the devotional meeting arranged by young Friends at the yearly meeting of "one branch" was really not a meeting of one branch, but of both. The responsible young Friends have a "tendency to draw together as unmistakable as was that of their ancestors to draw apart."

Friends in the Delaware Valley

One group of these young Friends had met fortnightly through one winter to study the foundations of Quaker faith as expressed by early Friends. Through the next winter the same group, their fellowship deepened, considered together such subjects as the nature of God, his revelation through Christ and through other sources, prayer, worship, service, and a final meeting in which each member told frankly what he considered the Dynamic of life. There were still among them differences of belief about "the Dynamic" but "complete acceptance of one another." In their third winter of association (1914-15) they sought to discover some fruits of spiritual life in social betterment. Some were very conservative, others radical, but among them there was great unity without uniformity.

One of the young men of this group, Henry J. Cadbury, wrote a long article entitled "Differences in Quaker Belief in 1827 and Today." Before the article was printed in the *Friends Intelligencer* each member of the group had expressed satisfaction with its content. The great interest in the article, when printed, called for its publication in pamphlet form.[25]

The remarkable thing about this group of young men and women is the active role its individual members played in all the activities of the two yearly meetings from 1915 onward to 1955, the eventual year of official reuniting. Their active roles very soon became leadership roles. Their convictions were deepened and their service heightened through the period of the First World War and its challenge both to the peace testimony and to the whole life.

As new generations of young Friends came into succession, and certainly through the next great challenge to a peace witness in the Second World War, the remarkable group of young Friends of 1913-1920 were wise and caring counselors and supporters of the new young Friends. Through the challenge and encouragement thus given, there was built up an even larger number of Friends in the Philadelphia Yearly Meetings who would not be satisfied until reuniting came. It was an exercise in patience, but also a time of great activity.

A few brief references to the individual contributions of certain

---

25. *Differences in Quaker Belief in 1827 and Today,* reprinted from *Friends Intelligencer,* The Biddle Press, Philadelphia, 1914.

young Friends will be helpful to the understanding of their roles as the movement toward reuniting of the yearly meetings progressed.

- The William Penn Lecture, sponsored by the Young Friends Movement, was given in 1916 by George A. Walton, still a young man. His topic: *The Quaker of the Coming Time.*
- Eliza Ambler (later Foulke), a secretary of the Young Friends Movement, wrote a *Friends Intelligencer* article in 1918, "The Quaker Message and the Young Friends Movement."
- A national Young Friends Conference held at Richmond, Indiana, in 1920 had six representatives from Race Street and seventeen from Arch Street. D. Herbert Way wrote the report for the *Intelligencer,* using one-fourth of his space to tell how much the Philadelphia delegation felt as one group and wished for the day they all could represent one yearly meeting in the Philadelphia area.

There were many others, including M. Albert Linton, William Eves III, Robert Pyle, Elizabeth Biddle (later Yarnall), Rebecca Biddle (later Bradbeer, then Alexander), Elizabeth A. Walters (later Furnas), Mary T. Sullivan (later Patterson), J. Barnard Walton, and D. Robert Yarnall. There were times of social recreation as well as study and worship and service. Caroline Phillips wrote in the *Intelligencer* in 1927 about a "Fun-Finders Group" at Wilmington, including members of the meetings of both branches, eight years before. The older members had come to do more serious things, "discussion of our responsibilities, and strengthening our lives by sharing our experiences."

In 1930 the official Young Friends Committee at Arch Street united with Young Friends at Race Street in "one organization called the Young Friends Movement, with two secretaries and an office at 1515 Cherry Street." In their report to the Yearly Meetings in 1932 the Young Friends Movement purpose was described as "to develop fuller confidence among Young Friends in the validity of the religious approach to the ultimate solution of world problems."

Continuing through the World War II period, the Young Friends Movement provided spiritual comradeship and encouragement, not only to young members of the Philadelphia Yearly Meetings, but also to young Friends and other conscientious objectors whose Civilian

Friends in the Delaware Valley

Public Service assignments or prisoner parole arrangements brought them to the Philadelphia area. Some young men in these groups became members of meetings in the area and made their contribution to the spirit of unity within these yearly meetings.

Other Influences Toward Unity

Elbert Russell, a Midwest Friend who had been made Director of the Woolman School at Swarthmore, wrote a series of articles in 1927 entitled "The Separation After a Century." He wrote, "It is significant and promising for the future progress toward unity, that progress has been most rapid where the two bodies know each other best."[26]

Many undertakings by Friends in Philadelphia were conscious efforts "to know each other." The Friends Historical Society and Friends Historical Association were merged in 1923. The Friends Social Union, limited to men and started in 1924, was organized on the assumption that men Friends had less opportunity than women to know one another socially. The establishment in 1929 of Pendle Hill, Quaker study center at Wallingford, was the work of a great array of most capable Friends from both yearly meetings.[27] Responsibility for much of the organizing of the second World Conference of Friends, and especially for serving as its hosts at Swarthmore and Haverford Colleges, was another experience in getting to know each other within the two yearly meetings.

From that World Conference came the beginnings of the Friends World Committee for Consultation, and the establishing of its American Section office in Philadelphia. The American Friends Fellowship Council, which originated within the American Friends Service Committee in 1935, for nearly fifteen years shared an office with the American Section of the FWCC; then the two organizations were merged. Philadelphia Friends had responsible roles in the development of these bodies which served Friends far beyond the borders

---

26. Elbert Russell, *The Separation after a Century,* 1928, reprinted from articles in *Friends Intelligencer,* October 1 to December 10, 1927.
27. Eleanore Price Mather, *Pendle Hill, A Quaker Experiment in Education and Community,* Pendle Hill, Wallingford, PA. 1980.

of their yearly meetings. They were similarly involved with establishing the Friends Committee on National Legislation whose office was set up in Washington in 1943.

Perhaps the closest association of Friends from the two branches came in the development of "united meetings" in local communities where there were two meetings, each having affiliation with a different yearly meeting. This movement began in the Germantown area of Philadelphia where the first "joint meeting" was held on April 18, 1920, and where some members of the Hicksite and of the Orthodox Meetings began to confer about union as early as 1924, then to meet together regularly. The result was an entirely new Chestnut Hill Meeting which became officially a United Monthly Meeting in 1933 when change in Disciplines of each yearly meeting made this status possible.[28] Radnor was the second United Monthly Meeting, with its founding members reopening a Friends Meeting House which had been unused for some time.

By 1955 when the two yearly meetings were reunited there were twenty-nine United Monthly Meetings and two United Quarterly Meetings.[29]

The periodical publications serving the separate yearly meetings during this period were *The Friend,* published by Managers who were members of the Arch Street Yearly Meeting, and the *Friends Intelligencer* which had the Hicksite Yearly Meetings as its clientele. *The Friend* was conservative in its news policy, printing straight information or actual documents representing Arch Street Yearly Meeting official actions and the activities of members and committees of that yearly meeting. Comment about those actions and activities appeared infrequently. The *Friends Intelligencer,* on the other hand, more freely and frequently expressed opinion about events and actions. In its pages there was frequent, and sometimes colorful, reporting of the events and the dynamics of both yearly meetings and of their members as they moved toward reunification.

Eventually the boards responsible for these two publications

---

28. Emily Cooper Johnson, *Under Quaker Appointment, The Life of Jane P. Rushmore,* University of Pennsylvania Press, 1953 (Chapter XI). Elizabeth Biddle Yarnall, *Chestnut Hill Monthly Meeeting* (c. 1972). In Quaker Collection, Haverford College Library, Haverford, Pa.

29. *Minutes,* Philadelphia Yearly Meeting (United), 1955, p. 73.

entered into negotiations to merge them into one periodical. These negotiations were conducted seriously in 1948 and early 1949. In the latter year it was concluded that the time had not arrived—that members of the two yearly meetings were not ready—for one periodical to serve both constituencies. The negotiations were dropped and Richard R. Wood, who had been part time editor of *The Friend,* was appointed full time "to strengthen *The Friend* editorially" in the hope of gaining additional readers and financial support. The *Friends Intelligencer* continued under the editorship of William Hubben.

The two publications developed in separate ways during the next five years, *The Friend* changing its format to pocket size. But the two sets of managers returned to negotiating of terms for merging the publications as the two Philadelphia yearly meetings moved quickly in 1954 toward reunification. At the Yearly Meetings in 1955 they announced the formation of Friends Publishing Corporation, an independent body, to be responsible for one periodical, wider in scope than the two former papers, with William Hubben as editor. On July 2, 1955, the new *Friends Journal* made its first appearance.

Toward More Certain Union

There were many actions by one or the other of the yearly meetings, or actions taken separately or jointly by both of them, which were positive steps—even if short ones—toward reunion. Some of them can be reviewed here. In 1926 Arch Street agreed to receive in their monthly meetings, members by certificate of transfer from any meeting in a yearly meeting of the Friends General Conference. Race Street responded with reciprocal action and recorded in their minutes, "The gradual passing away of the limitations which have separated us is a matter of deep thankfulness."

In 1928 each Yearly Meeting began the practice of appointing representatives to attend at least one session of the other meeting in that year, a practice which prompted expressions of appreciation. In 1930 the Race Street Meeting considered the suggestion from Arch Street that yearly meeting be held a few weeks earlier at Race Street so the two meetings might be held concurrently. Race Street turned down this suggestion, stating that the practice of visitor exchange which was proving valuable might be limited if both yearly meetings were in session at the same time.

Charles F. Jenkins, a Race Street Friend from Germantown, had written a pamphlet which proposed a specific plan for uniting the two yearly meetings. Race Street in 1932 received the pamphlet favorably and referred it to Representative Committee for further consideration.[30]

By 1940 the two Yearly Meetings were meeting concurrently at their respective meeting houses. In that year one session was held jointly when "in view of the world crisis" consideration was given to the subject of the peace testimony. The number of joint sessions was increased to two or three in subsequent years, and use made of those sessions to hear reports from the AFSC or the FWCC or other organizations to which both yearly meetings were affiliated. The Young Friends Movement soon asked to be permitted to report at a joint session, and this was approved. Requests from joint committees to be allowed to report at joint sessions were not favored, however, unless each yearly meeting in separate session agreed to the request. But by 1943 there had been a revision of Discipline which authorized the yearly meetings in joint session to make decisions on behalf of each yearly meeting.

The decision to hold evening sessions, and to forego morning sessions on weekdays, came in 1944, to make possible the participation of a larger number of young Friends. In 1945 a pamphlet entitled "A Plea for Unity" was prepared by the Joint Committee of the Society of Friends for Montgomery and Bucks Counties. The writers of this pamphlet urged that the two Philadelphia Yearly Meetings unite at once. They suggested frequent reminders that "Friends might carry more weight in their testimonies before the world if they made peace among themselves by abandoning the feud of over a century." Race Street Yearly Meeting referred the pamphlet and its message to Representative Committee with instruction "to explore all avenues through which closer cooperation may be established." Arch Street Meeting would be informed of this action with the hope that it, too, might be willing to refer this matter to a similar committee. Arch

---

30. *Minutes,* PYM (Race Street), 1932, pp. 4–5.

Street "warmly endorsed the goal of essential unity in spirit and action, not only of the two Philadelphia Yearly Meetings but of all Friends in America." They referred the whole matter to Representative Meeting, directing that body to cooperate with Race Street in accordance with action already taken there.[31]

The following year, 1946, a plan for moving closer to reunion was proposed. There would be established a "General Meeting" comprised of all members of both Yearly Meetings, to gather at a time other than the holding of the two Yearly Meetings which would continue to be autonomous bodies. Both Yearly Meetings accepted this proposal and each named twenty-five members to a planning committee for the General Meeting. The function of the General Meeting, in brief, would be (1) to convene in autumn each year to consider the spiritual and temporal concerns of Friends of the Philadelphia area, but without legislative powers except as granted by concurrent action of both meetings, and (2) to receive committee and organization reports in which the yearly meetings have common concern.

The General Meeting and—Reunion At Last

The Philadelphia General Meeting, then, became the instrument which Friends hoped would enable their further progress to full unity. It first met on November 1 and 2, 1946. M. Albert Linton was the Presiding Clerk, with Eleanor Stabler Clarke serving as Recording Clerk. Steps were taken to assure participation by young Friends in the Planning Committee, and George Walton, on behalf of the Planning Committee, made a superb introductory statement which enumerated four goals to be achieved through the General Meeting:[32]

1. Among members there should be a feeling of increasing identity with the General Meeting—a single body—rather than with the separate yearly meetings. This would not be the transfer of an old loyalty, but creation of a new one.

---

31. *Minutes,* PYM (Arch Street), 1945, pp. 19-20. *Minutes,* PYM (Race Street), 1945, pp. 16-18. *The Friend,* April 12, 1945, pp. 328-329. *Friends Intelligencer,* February 17, 1945, pp. 108-110.
32. *Minutes,* Philadelphia General Meeting of Friends, 1946, p. 7.

Diminishing Separation

2. Increasing public recognition that Friends in this area are one body is desired.
3. Growing out of 1 and 2 above, greater influence and power in furthering our testimonies should be realized.
4. There should develop a deeper spiritual experience and an enhanced sense of fellowship growing out of belonging one to another.

There was ample time given to worship and to assessment of the spiritual condition of Philadelphia Friends in the first General Meeting and in those which followed. The kind of reports which a few years earlier were heard in joint sessions of the two Yearly Meetings were now given at General Meeting. There was a sense of "new things" in this arrangement, and response of members was enthusiastic. The first General Meeting was limited to two days.

It was not easy to turn back requests from a variety of yearly meeting joint committees to be placed on the General Meeting agenda "next time." But the Planning Committee wisely resisted many such requests, wanting to make this something "new" and not a mere repetition of old experience.

George Walton was made General Meeting Clerk after the second year, followed in a succession of one-year terms by Anna Cox Brinton, Clarence E. Pickett, Caroline N. Jacob, with Charles Darlington and James F. Walker Co-Clerks in the final year, 1954.

In 1950 the General Meeting Planning Committee brought to the yearly meetings a proposal that each Yearly Meeting appoint members to a new committee to prepare a common Book of Discipline. It suggested that the beliefs and practices in the two yearly meetings were not highly divergent, and that a synthesized Discipline might be another unifying influence. The confidence of both yearly meetings in the leadership of George Walton is evident in their appointing him to be chairman of the Committee to Prepare a Common Book of Discipline. This task was soon to be shared between two parts of the original committee. The Committee to Prepare a Common Book of Discipline retained responsibility for Part I of the new book, with focus on introductory and historical material describing Quakerism, and Part III which presented the thought and faith of Friends through quotations gleaned from the record of this religious society's three-hundred year history.

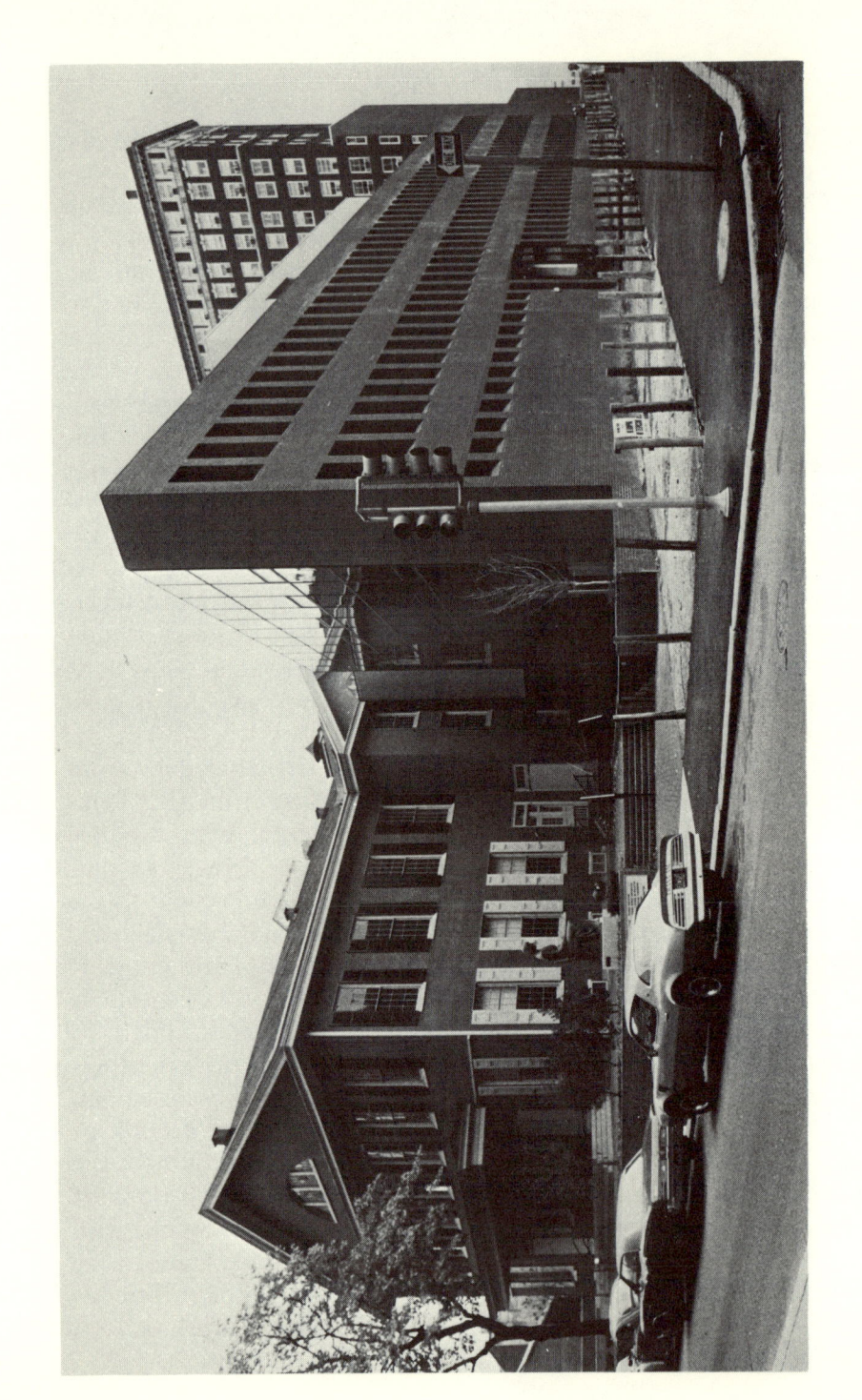

Friends Center, Fifteenth and Cherry Streets, 1975. Photo by Theodore B. Hetzel.

Diminishing Separation

James F. Walker was made chairman of the companion committee, set off from the Committee to Prepare a Common Book of Discipline—the Committee on Organic Union. It had the Responsibility for drafting Part II of what was later given the title *A Book of Faith and Practice*. Part II dealt with the practical arrangements for one yearly meeting which would be the result of the merger of two existing yearly meetings in the Philadelphia area.

It is frequently said that one incident, more than any other, made possible the final decision by many in both yearly meetings to set their feet firmly on the path to reunion. Thomas and Eliza Ambler Foulke had taken leave of their Philadelphia responsibilities and had given one year of service for Friends in Japan. There they worked in many ways very closely with Japanese Friends for whom they had great respect and affection. They soon came to realize that, for Friends in Japan, a division of the Society of Friends into two bodies unable to unite was incomprehensible. Thomas and Eliza, therefore, returned from Japan to Philadelphia very much under the burden of concern that Philadelphia Friends must find the way to put aside the hindrances to reunion and to make more rapid progress to that goal. They were members of the Race Street Yearly Meeting and they first shared their concern there in the Yearly Meeting of 1950. The response was in full sympathy with the Foulkes' concern, and the yearly meeting minute authorized, if Arch Street Meeting should concur, the Planning Committee of General Meeting to proceed in such a way as to realize complete merger of the two Philadelphia Yearly Meetings in the following year. If Arch Street should prefer it, it would be acceptable for the two Representative bodies to take this matter in hand and move as fast as possible toward achievement.

Arch Street also responded positively but with greater caution: "The fact that we want to go forward does not mean that the process is entirely easy. There are practical problems. . . for example, the care of property and the management of our schools. We shall do well to think of the difficulties as soluble problems." The concern was referred to the two Representative bodies for study, and to confer together, looking toward the ways in which the union of the two yearly meetings could be accomplished.

During the years 1951–1954, the Committee on Organic Union and the Committee to Prepare a Common Discipline (now working on

Friends in the Delaware Valley

the Section called Faith and Practice) worked diligently. By 1954 a book on Faith and Practice had been prepared, but the Committee recognized that it was a document in transition, that more work must be done after the act of reuniting was taken and Friends had this experience behind them. The Committee on Organic Unity had produced a book of Church Government which had been used for a year on trial basis. Suggestions, although not many of significance, had been received, had been considered and incorporated in a revised text if the committee felt a suggestion was justified. These reports were made to the two Yearly Meetings in the Spring of 1954 and to General Meeting in the autumn. Details for further attention were noted and all was made ready for the Yearly Meeting sessions of 1955.

On the 26th of March 1955—a Seventh Day afternoon—the Philadelphia Yearly Meeting at Arch Street and the Philadelphia Yearly Meeting at Race Street, meeting separately, approved their reuniting to become one body, the Philadelphia Yearly Meeting of the Religious Society of Friends. One hundred and twenty-eight years of separation were ended!

The reunited Philadelphia Yearly Meeting, bringing into membership together 4,479 members from Arch Street and 10,793 from Race Street, met as one body on Second Day morning, March 28, 1955. The duties of Presiding Clerk were shared by Charles Darlington and James Walker. The opening minute was a prayer:

> We are born again—a new Yearly Meeting. We ask to be sustained by the bread of life and the water of life. We ask to be guided by our Heavenly Father's counsel, not only in this week but in the adjustments that lie ahead.

V A Widening Path:
Women in the Philadelphia Yearly Meeting Move Toward Equality, 1681–1929

by MARGARET HOPE BACON

The Quaker concept of the continuing revelation of Truth is well illustrated in the development of a concern for the equality of women. Although the founders of the Society of Friends believed in women's spiritual equality—their right to preach and prophesy and play a role in the meeting for worship—they did not see the implications of this insight for secular life. It has taken over three hundred years, in fact, for a fully developed testimony of sexual equality, within and without the Society, to develop.

To read the minutes of the Philadelphia Yearly Meeting of Women from 1681, when the first entry was made, to 1929 when the Orthodox wing appointed its last women's clerk, is to watch the slow and uneven growth of the women's sense of the role they must assume in order to become full partners—Fox would have called them helpsmeet—in the work of the church. At times the women's attitude seemed to be that of timidity and even undue deference to their brothers, and progress was painfully slow. Yet taken as a whole, over the 248 years, the story is an inspiring one, parallel to that of the slow development of a full-blown testimony against discrimination on the basis of race.

Friends in the Delaware Valley

The Philadelphia Yearly Meeting of Women Friends was established approximately twenty-nine years after the Society of Friends was born. In that intervening period much had already happened to determine the shape of woman's place in the meeting community.

In common with other radical protestant groups of the period, Fox and his early followers defended woman's role to preach and prophesy. Both George Fox and Margaret Fell wrote books on the subject.[1] Quaker women feeling divinely led to travel in the ministry were encouraged to do so. Fox's first convert, Elizabeth Hooten, a grandmother, travelled up and down England and in the colonies. Two women, Ann Austin and Mary Fisher, brought the Quaker message to Boston, while Elizabeth Harris planted the seed in Virginia. The latitude allowed the travelling women ministers, their courage and their exploits, is a remarkable aspect of early Quaker history.

In 1656, while the Quaker movement was yet very new, Fox was inspired to see if these valiant Quaker women could not also play a role in church government. He first encouraged a group of women to establish the "Box Meeting," to provide for the poor and needy. Shortly thereafter he helped in the creation of a Two Weeks Meeting with the purpose of sending women to visit the sick and the prisoners and to look after the poor, the widowed, and the orphaned. This meeting was particularly useful in the period of persecution of the Quakers.

The success of these two meetings evidently further encouraged Fox in his belief in women's abilities. In 1671, shortly after the establishment of regular monthly and quarterly meetings for business, he began to advocate the setting up of parallel meetings for women to attend to "women's matters." In this he was opposed by two prominent leaders of the new Quaker movement, John Story and John Wilkinson, who objected that it was adding an unnecessary degree of organization to the Society. Fox responded with his usual vigor:

---

1. In 1656 George Fox published *The Woman Learning in Silence, or the Mysterie of Woman's Subjection to her Husband, as also, the Daughters Prophesying Wherein the Lord hath, and is Fulfilling that he Spake by the Prophet Joel, I will pour out my Spirit unto All Flesh.* In 1661, he published a second tract, *Concerning Sons and Daughters, and Prophetesses Speaking and Prophesying in the Law and the Gospel.* In 1666 Margaret Fell published her first work on the subject, *Women's Speaking Justified, Proved, and Allowed of by the Scriptures, all such as Speak in the Spirit and Power of the Lord Jesus.*

A Widening Path

> There is some dark spirits that would have no women's meetings, but as men should meet with them, which women cannot for civility and modesty's sake speak amongst men of women's matters, neither can modest men desire it, and none but Ranters will desire to look into women's matters.[2]

William Penn in *Just Measures,* written in 1692, took the same view:

> But it is asked why should women meet apart? We think for a very good reason. The church increaseth, which increaseth the business of the church, and women, whose bashfulness will not permit them to say or do much, as to church affairs before men, when by themselves, may exercise their gift of wisdom and understanding, in a direct care of their own sex, at least, which makes up not the least part of the business of the church, and this, while the men are upon their own proper business, also, as men and women make up the church, men and women make up the business of the church.

Other schismatics took issue with the separate women's meetings, but Fox, Margaret Fell, and her daughters continued to defend them vigorously, and urged their establishment in the New World. In 1676 Fox wrote a long epistle, "Encouragement to all the Women's Meetings in the World" which reached the American colonies. At about the same time Sarah Fell, Margaret's daughter, clerk of Lancashire Quarterly Meeting for Women, wrote an epistle to "Women's Meetings Everywhere," urging them to keep good records and to meet regularly.

Despite this strong early support for women's meetings, the London Yearly Meeting did not establish a Yearly Meeting for Women until 1784, and then after a good deal of pressure from the Americans. Women came to London during Yearly Meeting time, but simply met for worship in various meeting houses in the city while the men conducted the business. Dublin, however, developed a separate women's meeting in 1679, and the various yearly meetings that developed in the New World each had a parallel women's body. It may be that the general air of liberty, and the frontier conditions of colonial life, which demanded that husband and wife share responsibility for the home, the farm, or the shop, had something to do with this development.

---

2. George Fox's *Epistles,* No. 313 as quoted in W. C. Braithwaite, *The Second Period of Quakerism,* p. 274, Note 4.

Friends in the Delaware Valley

Philadelphia Yearly Meeting for Women: The Early Years, 1681-1755

Philadelphia Yearly Meeting began with a general meeting in Burlington, New Jersey on September 10, 1681 (N.S.). The first entry of the Women's Yearly Meeting was made at this time, and reads: "At a general meeting held in Burlington the last of sixth month, 1681 at the house of Thomas Gardner, it was mutually agreed that a Women's Meeting should be established."

The next entry in the women's Yearly Meeting minute book is dated Seventh Month 9, 1691 and reads: "Pursuant to the above-mentioned establishment an annual meeting of women Friends was from time to time held in the Service of Truth of which a record was made and being left sometimes in the hands of one Friend and sometimes in that of another, the record serving both the Provinces by which means it is not now to be found and we are at a loss for the minutes of the several years between 1681 and 1691."

Women Friends in England had prided themselves on their care in keeping minutes. The same attitude prevailed early in the colonies. The selection of a member to keep all the meeting papers in order early became one of the important functions of the women's Yearly Meeting. This early breach was therefore bitterly regretted. In 1784 another explanation for the missing minutes was entered: "That the minutes of the first ten years after the establishment of the meeting being supposed to be in the hands of a person who joined in that unhappy separation promoted by George Keith they could never be restored, which information was satisfactory to the meeting."

Whatever the reason for the hiatus, the 1691 meeting made up for lost time. It developed a system of hearing reports from the representatives from each quarterly meeting. It urged all meetings to take care in their dealings with the Indians, particularly not to give them liquor. And it asked that each monthly meeting take up a collection and send it to the nearest quarterly meeting that it might be sent as a token of love to women Friends in England.

Following this first burst of activity, the next decade of women's Yearly Meeting appears uneventful. One gains the impression that Yearly Meeting women were sometimes at a loss as to what business to conduct. As the representative of each quarterly meeting was called to report, the reply was often rather short: "Salem Quarter has nothing to report, but sends love to Friends." In 1698 the minutes of two travelling women Friends were read and accepted, a sign that the travelling ministry was still vigorous. In 1699, the first of many epistles was prepared to be sent to women Friends in London.

The exchange of epistles with corresponding women's yearly meetings became an important aspect of women's Yearly Meeting. The reading of an epistle from another meeting, the decision to answer it, the appointment of a committee to prepare an answer, the reading of the answer to the group, and its final copying in a fair hand and mailing were all handled with due weight, and provided the women with valuable experience. After five years of writing to the women in London, the Philadelphia group began to correspond with Maryland in 1704.

Disturbed by the Keithian controversy, the early years of Philadelphia Yearly Meeting were tumultuous. Aside from the loss of the minutes some suggestion of the trouble is contained in a minute in the women's Yearly Meeting of 1703: "It is likewise requested and desired by this Meeting that for time to come Friends belonging to it shall not openly oppose each other before being publickly disowned amongst them."

Nevertheless, the women's Yearly Meeting was continuing to move ahead. In 1705 the women took a giant step forward by deciding to make an annual collection through the women's monthly and quarterly business meetings, in order to possess a yearly meeting fund, or "stock," as they called it, of their own. They would not, however, be able to do so unless this met with "the approbation and concurrence of men Friends." The timidity with which the women approached this request to the men's Yearly Meeting, and the somewhat lordly attitude with which it was granted is reflected in the minute from the men's Yearly Meeting:

> Samuel Jennings proposed, on behalf and at the request of the women's Meeting that they may be permitted to have a Yearly Meeting stock of their own, for such services as may properly fall under their notice, to which this meeting condescends.
>
> Signed, Isaac Norris

Nevertheless, having a separate fund immediately gave the women's meeting more responsibility. It was now necessary to keep accounts, to elect annually a treasurer and her assistant, and to decide each year what to do with the money. In 1706 women Friends decided to give their first collection to "dear friends and sisters in East New England who are suffering great distress."

Perhaps emboldened by the acquisition of their own funds, the women of Philadelphia Yearly Meeting at this time ventured to send their first message to the men's Yearly Meeting. The message recommended that the book of discipline be duly observed and put into practice in all monthly and quarterly meetings and reported on at Yearly Meeting; a practice that was not standardized until some fifty years later.

As the early years of the eighteenth century passed, Philadelphia Yearly Meeting of Women began to increase the number of epistles they sent to other Quaker women. New England (variously listed as Rhode Island or Newport) was added in 1708; Long Island (Flushing) in 1746, and Friends in Virginia and North Carolina in 1747. The letters, minutes and accounts of the meeting were already so weighty by 1728 that a member, Beulah Coates, was chosen to obtain a suitable trunk to keep them all in. In this year, too, a more formal method of selecting a clerk each year was begun, the representatives from the various Quarters retiring early in the meeting, then reporting their choice. This system was carefully followed thereafter.

A concern for the morals and education of youth became a theme of the women's Yearly Meeting in its early days. In 1714 an epistle of advice to young Friends was prepared:

> Many tender Friends having expressed their great grief and sorrow of some particular things too much of late crept in amongst Friends children, as parting the hair, pinching the caps around the face, pleated and leaded sleeves, bare backs and breasts with gay stomachers all which this meeting tenderly advises our young people to keep out of, remembering how beautiful the King's daughter of old was, who was all glories within, to which ornament with that of a meek and quiet spirit which is the Light of God is of great price.

Four women were chosen to convey this epistle in person to the men's Yearly Meeting. This is the first time such an action is recorded in the minute books.

In similar vein, in 1728 the women's Yearly Meeting warned Friends to maintain the testimony "of our ancient Friends against wearing black or black and white cloathes at Burials." In 1739 the wearing of hoop skirts, and the use of fancy fans, was deplored.

A Widening Path

Having aided suffering Friends in New England, Philadelphia Yearly Meeting women began to use their stock, the money they collected each year, "for the necessities of those who go to sea" or those who travel in the ministry, whether male or female.

Years of Withdrawal: 1755–1827

The year 1755 was a turning point for Philadelphia Yearly Meeting as a whole, as recorded elsewhere in this volume, a turning away from the affairs of the world and into the affairs of the Society. The regular reading and answering of queries was suggested in 1755 and begun in 1756. Thereafter for many years questions of discipline were uppermost in Yearly Meeting minutes. Letters of advice were regularly sent to quarterly and monthly meetings, and committees of men and women Friends were set up to visit all families within the Yearly Meeting. There was increasing concern about the institution of marriage. Friends were regularly warned against a "drowsy spirit," and as time passed there was more and more emphasis on giving Friends' children a religiously guarded education.

As the men Friends turned away from the world and into the church, they appeared to take more interest in women's Yearly Meeting. In both 1756 and 1757 men Friends appeared before the women's meeting with "seasonable advice and counsel to communicate." Thereafter such visits occurred frequently. In 1756 the women appointed a committee to overhaul the Queries and to "contract them by omitting such parts as appear calculated principally for the service of men Friends." The men, however, instructed the women to answer the Queries as written, omitting the word "brethren" and inserting "followers of Christ" when that was appropriate.

Although the women's minutes continue to reveal a rather docile spirit, the women took the initiative at this time on two important matters. The first concerned the establishment of a Yearly Meeting for Women Friends in London. In 1762 the Philadelphia women began the practice of copying the epistle they received annually from the London Friends, and sending it to all quarterly meetings to be read. They evidently thought the same practice would be followed in England. When they learned that this was not done, because the group that met informally in London did not consist of representatives from the quarters, they resolved to try to remedy the situation.

Friends in the Delaware Valley

In 1766 they minuted their action:

> Divers well concerned Friends of this Meeting being under a weighty sense of the great disadvantage our Worthy Women Friends in London labor under for want of a yearly meeting properly established amongst them for the discipline of the church, have agreed to solicit the mens meeting to take the same under their consideration in order to propose in what manner they shall think best to Friends in London that they may be favored with the same privilege we are in this respect. For this service, the Meeting appoints said Friends: Ann Moore, Susanna Lightfoot, Esther White and Margaret Churchman, to attend the Men's Meeting.

And later in the minutes:

> The friends appointed to attend the Men's Meeting with the above Minute returned with an answer in writing signifying they received the proposal of a Yearly Meeting for Women Friends in London as a matter of weight, and should give it due consideration, which was likewise confirmed by a message from the Men's Meeting delivered by three Friends, viz: John Churchman, Isaac Child, and Thomas Milhouse to this purpose that they truly sympathized with the present circumstances of our Friends in England, and that they had appointed a committee to consider the affair, who upon solidly deliberating they agreed to report that it was their sense and judgment that Friends here should abide under the weight of the concern and exercise, until a more convenient time offered to move thereon, especially as they had been informed that Friends in England had lately endeavored to bring about such a work, without the desired effect.

As a meeting, Philadelphia Friends made no further efforts to remedy the situation. Several men and women were, however, involved in keeping the concern alive. In 1784 when Rebecca Jones, a member of the Yearly Meeting, traveled to England, she was able to play a supportive role to the English women who finally obtained permission to have their own meeting.

The second initiative of the women during this period concerned their place of gathering. Up until 1760 Philadelphia Yearly Meeting as a whole alternated between Burlington one year and Philadelphia the next. In Philadelphia the Great Meeting House on the southwest corner of Second and Market was used for these annual gatherings. It was not large enough to accommodate both the men and the women's yearly meetings, so either one or the other group adjourned to the Bank Street meeting house, on Front Street north of Arch. In 1755 the Great Meeting House was rebuilt on its existing site, and was known as the Greater Meeting House. The men now held their

A Widening Path

annual meeting at the new Pine Street Meeting House, so the women met in the Greater Meeting House each year from 1760 to 1804.

This meeting house had one great disadvantage, that of noise. On every day but First Day the market stalls, or shambles along Market Street, were in full cry. Under these conditions, the women found it difficult to hear. In 1794 the women resolved to take their concern to the men's Yearly Meeting. A large committee, including the weighty Rebecca Jones, was selected for this task.

This committee reported back in 1795 that a more suitable place, the burying ground at Fourth and Mulberry, was under consideration, but that the men did not regard it expedient to erect a building. After some discussion, the women minuted their unresolved concern: "This meeting believes it right to inform Men friends they have been more sensible of the necessity of providing a more convenient place to meet at this time. We desire to submit and yet hope and much desire that way will open."

Again in 1798 the Women's Yearly Meeting appointed several members to voice their concern about the meeting house. Finally in 1803 they were able to record that the men had agreed to begin construction at Fourth and Mulberry (now Arch). The women's Yearly Meeting met in what we now call the East Room until 1811, when the West Room was built. Thereafter the women moved to the West, and the men's Yearly Meeting, which had been using the Northern District Meeting at the Key's Alley Meeting House, met in the East Room.

The second half of the eighteenth century was a time of change for all Friends. The pioneer generation was dying off. In 1763 the women's Yearly Meeting recorded a memorial brought forth by Haddonfield Monthly Meeting for Elizabeth Estaugh who died at the age of 82. As Elizabeth Haddon, she had come to West Jersey in 1701 to take charge of her father's land, and to preach. A year after her arrival she met and married John Estaugh, a travelling Quaker minister. Theirs was a marriage of equals, unusual in the colonial period.[3]

A new generation of Quaker women was beginning to travel in

---

3. There are numerous sources for material on Elizabeth Estaugh. The best is an unpublished manuscript in the Frank H. Stewart Collection of Glassboro (New Jersey) State Teachers College.

Friends in the Delaware Valley

the ministry. During this period Philadelphia had visits from Susanna Hutton of Dublin, Rachel Willet of Long Island, Charity Cook of North Carolina, and Lydia Rotch of Nantucket, among others. Philadelphia also sent one of their members, Sarah Nicholson, to visit among Friends in the Carolinas.

A concern against slavery, which was beginning to affect the whole Yearly Meeting, was discussed by the women. In 1763 it was noted that some Friends were not as careful as they should be in their education of their slaves. Within a few years women were beginning to report on the freedom of members of the various quarters from buying or importing slaves. In 1776 Philadelphia Yearly Meeting ruled that it would be a disownable offense to keep slaves. The women succinctly reported, "Many have set their slaves at liberty since last year." In 1784, women were appointed to a committee to work with men in assisting with the care and oversight of free Negroes.

An entry in the diary of Elizabeth Drinker, a Quaker minister and member of Philadelphia Yearly Meeting, reveals how quickly sentiment had changed in regard to slavery. In 1807 she recorded that she had visited a slave whom she had sold as a child of eight fifty-one years before, in 1756. "When we sold her, there was nothing said against keeping or selling negroes." Some years later they repented and tried to repurchase her, but it was not until the slave was grown that "there was much talk of the inequity of holding them in bondage."[4] In other words, the prominent and well-meaning Drinkers apparently did not consider slavery wrong until about 1770.

Concurrent with the Meeting's decision to rid members of slaveholding was the beginning of the American Revolution. In the state of retirement from the world in which Philadelphia Friends had been since 1755, they allowed the "curent time of tryal and outward commotion" to make very little change in the orderly procedures of their Yearly Meetings. A group of men Friends visited the women's Meeting in 1775, "expressing their affliction for the late deviations of many making the profession to the truth who are not adhering thereto, having been concerned in making preparations and exercising themselves for war, and receiving paper bills of credit." Friends from London wrote offering sympathy and advice. The women

---

4. *Extracts from the Journal of Elizabeth Drinker,* Henry D. Biddle (ed.) Philadelphia: Lippincott, 1889, p. 409.

A Widening Path

recorded a memorial for John Hunt, exiled to Winchester, Virginia as disloyal after American troops occupied Philadelphia, who died while in exile.

Perhaps as a result of the fact that some young firebrands forgot their Quaker pacifism and went to war, there were frequent references in the minutes of the period to the need to raise sons in the service of truth, and to the desirability of a religiously guarded education. In 1794 the women's Meeting minuted its hope "that our dear young women may seriously consider which of them are called upon to dedicate their time and talents to the Lord, by endeavoring to resign themselves to the useful occupation of instructing children in the profitable parts of school learning." In 1795 a group of women were appointed to a committee to study the feasibility of a boarding school. In 1799 it was recorded that the school, "West Town," would begin the next month with twenty pupils of each sex, "divers Friends have acceptably surrendered their names for the different stations of the family." ("The family" was the name given the school staff.)

Although boys and girls were kept strictly apart, Westtown was coeducational to the extent it offered both sexes a similar curriculum; the boys, however, had the opportunity to study Latin and Greek, while girls were required to study sewing. Perhaps more important, the fact that both men and women Friends served on the School Committee provided women with a chance to exercise their talents, although their original concerns were primarily with housekeeping and the morals of the girls.

Westtown grew rapidly. By 1800 there were one hundred boys and a like number of girls enrolled, at a cost of thirty pounds a year. Each year thereafter a reading of the school committee report was an integral part of women's Yearly Meeting.[5]

In 1800 the Yearly Meeting had an application from Pelham Monthly Meeting in Ontario to become part of Philadelphia Yearly Meeting. (It later joined New York Yearly Meeting.) In considering Pelham's application it was discovered that men and women had been meeting together for business all along. Philadelphia's first answer to Pelham was that men and women must sit separately for business. The next year, however, this stipulation was withdrawn.

---

5. Helen G. Hole, *Westtown Through the Years,* 1799-1942, Westtown Alumni Assoc., 1942.

"Men and women should sit together whilst conducting business as heretofore."

As Friends began to move westward the application of new groups wanting to establish themselves as quarterly or yearly meetings began to appear before Philadelphia. In 1820 and 1821 the establishment of Indiana Yearly Meeting at White Water came before the women's meeting as well as the men's. By 1824 the women were prepared to exchange epistles with both Indiana and Ohio Yearly Meeting.

Parallel with the western migration was an increasing concern for the American Indians. In 1795 a committee of men Friends first brought a concern for the Indians to women's Yearly Meeting. In 1801 the idea of establishing a boarding school for Indians was discussed. In 1804 the women's Yearly Meeting minuted its hope that one or more women Friends would be prepared to go and assist at the new Indian school at Tunesassa, helping with "the dairies, spinning, and other domestic economies." Perhaps a couple and one single woman would be the Tunesassa "family," it was suggested. Women Friends, however, were not invited to serve on the Indian committee.

We know from the letters and diaries of the period that the stresses and strains leading up to the Hicksite-Orthodox separation were affecting the Yearly Meeting in the first quarter of the nineteenth century. Little of the turmoil, however, found its way into the serene conduct of women's Yearly Meeting. In 1824 Anna Braithwaite, a British minister, was recorded as present at the Yearly Meeting. Again in 1826 another British visitor, Elizabeth Robson, was in attendance. Both had been sent to see what could be done to stem the discord, and to insure the soundness of doctrine preached in meeting. At the crucial meeting in 1827, when the two factions could not agree on a clerk in the men's meeting, and the Hicksites withdrew, the only sign of conflict in the women's minutes is one terse statement: "Under a feeling of exercise on account of the state of our religious society we believe it right to appoint a committee to visit all Quarterly and Monthly Meetings."

This minute was taken by several women to the men's Yearly Meeting. For Orthodox women, it was the last initiative they were to take for almost ninety years.

A Widening Path

The Orthodox Women's Yearly Meeting Moves Slowly: 1827–1929

From 1827 until 1922 there were two sets of Yearly Meeting women's minutes—the Hicksite and the Orthodox. The women's minutes came to an end when both Orthodox and Hicksite decided at different times to hold integrated male/female meetings. It was not until 1955 that the two branches rejoined. To read these ninety-five years of parallel minutes is to be struck by a real differences in tone and style between the two sets of women. The Hicksite women were more assertive, initiating minutes and undertaking concerns without waiting for the prior approval of the men. They served on more joint committees, began agitating earlier for an equal role in the discipline of their own members, and ultimately began to work for, and achieved representation on the powerful Representative Meeting and thus an equal role in church governance forty years before the Orthodox were so represented.

Various guesses can be made about this striking difference. The Orthodox were more urban, more well-to-do, more conforming to the mores of the larger community. They may have absorbed some of the strong feelings about women's proper sphere so prevalent among the middle and upper middle class in the nineteenth century. The Orthodox were strongly influenced by the Evangelicals, many of whom opposed the reform movements of the day—whether temperance, abolition or women's rights—as a distraction from, if not a substitute for, a devotion to Christ. (There even came to be an Evangelical wing within the Hicksite group with this point of view.) Although not all Hicksites were liberals, the position they had originally taken in advocating freedom of expression in meeting pushed them toward more democratic views. Finally the Hicksites had a number of strong women, one of whom, Lucretia Mott, advocated both the abolition of slavery and the rights of women both within the Hicksite Yearly Meetings and the larger community.

None of this seems to account for the very humble, almost obsequious tone the Orthodox Yearly Meeting women sometimes took toward their male counterparts. We know from their journals and letters that many of them were strong women and vigorous ministers. One can only note that something of this tone can be found in the London Yearly Meeting women's minutes of the same period, and suppose the tone reflected the manners of the society of the day, more than the Society of Friends.

In 1828 and 1829 the minutes of the Yearly Meeting of Women Friends (Orthodox) were full of reports of meetings being set up "not in unity" with Friends, and statements lamenting the spirit of libertinism and insubordination. There were reports of other separations in other Yearly Meetings—Baltimore, Indiana, New York, and Ohio. The women were comforted by the return visit of Anna Braithwaite in 1828, and supportive epistles from London and from Ireland.

By 1830, the Yearly Meeting women seemed to have accepted the fact of the separation as a *fait accompli*. The problem, the women minuted in 1831, was the fact that the Separatists had retained control of many of the school houses, and it would be necessary to raise money in order to provide Quaker youth with a religiously guarded education. Westtown had remained in Orthodox hands, and the report of the Westtown committee was an important element of each women's Yearly Meeting. They also heard each year from the committee overseeing the settlement and school at Tunesassa.

Through the 1830's Orthodox women Friends went with men Friends to visit all subordinate meetings. The Hicksites did the same. Both groups were trying to win converts and consolidate gains. The Orthodox continued to have British Friends visiting them and helping them in this work. Joseph John Gurney came in 1838, Elizabeth Robson in 1839, and Daniel Wheeler in the same year.

In 1842 a new practice began at women's Yearly Meeting. The minutes of the all male Meeting for Sufferings (later Representative Meeting) were read to the women in their regular sessions. The women felt that this was a great privilege. In 1843 they minuted: "We have been deeply interested in being thus permitted by men friends to unite in feeling with our beloved brethren on the various subjects of both civil and religious importance that have from time to time engaged their attention."

By 1845 a second schism, betweeen the Gurneyites and Wilburites, was under way, unhappily affecting the Orthodox Yearly Meeting in Philadelphia. In 1846 the women, at the suggestion of men Friends, gave up correspondence with New England Yearly Meeting, where

A Widening Path

the schism had originated. Thereafter correspondence with other affected yearly meetings was given up, until by 1857 Philadelphia was completely isolated.

Along with keeping accounts and writing minutes, the yearly job of writing epistles to sister Yearly Meetings had been one of the major functions of the women's Yearly Meeting since its inception. The loss of this task, which had first given the women confidence in themselves as a meeting, may have had something to do with the extremely quiet condition into which the women's Yearly Meeting subsequently sank. Reading the minutes from 1857 to approximately 1887, one senses a gradual loss of remaining effectiveness and power.

One aspect of the quietness was the distance between the women's Yearly Meeting and the events of the day. The concern about slavery which had been prominent earlier was felt to be too exciting, and its mention was simply ruled out of the Yearly Meeting. There was occasional mention of the Indian problems, but the primary focus continued to be the school in Tunesassa. There was no mention of the war of 1846 in the women's minutes and the Civil War was mentioned only briefly in the report of the Meeting for Sufferings in 1862 and 1864.

Outside of the structure of the Yearly Meeting, however, the Orthodox women united during the Civil War to pack clothes for the "contraband," the black men and women in areas under the control of the Union Army. They began this enterprise in 1862.[6] Not until the end of 1863 did the Orthodox men form an association, the Friends Freedmen's Association, for this purpose. The women continued to pack, but they were not invited to attend the meetings of the new group.

Stories of the sufferings of Friends in the South were reaching Philadelphia, and for several years the Orthodox women used their funds for the relief of such sufferings, first in Tennessee and then in North Carolina. Later, some of these funds were sent to Kansas, where Friends such as Elizabeth Comstock were helping destitute blacks. After these calls on the exchequer were settled, the women

---

6. Lucretia Mott to Martha Coffin Wright, Dec. 27, 1862. Mott Mss., Friends Historical Library, Swarthmore College.

decided to spend their money on providing simple meals for Friends during Yearly Meeting, and this practice was continued into the twentieth century.

Midwestern women Friends were becoming interested in missionary work. In September of 1881 a Women's Foreign Missionary Society was formed in Indiana. The next year a Women Friends Missionary Society was formed by Orthodox women in Philadelphia. Meeting with some Japanese women in 1885, the group decided to establish a mission in Japan, and a school was subsequently opened in 1887. Two years later men were admitted to membership. All of this was outside of the Yearly Meeting. When a large group of Quaker women from the various Yearly Meetings met at Glens Falls, New York in 1890 to form the group that is now the United Society of Friends Women, several Philadelphia women were present, but not as meeting representatives. It was not until 1923 that the group, renamed the Foreign Missionary Association, began to report to Yearly Meeting.[7]

In 1877, the year Hicksite women won an equal role in their Yearly Meeting, the Orthodox women still sat on only one Yearly Meeting committee, the Westtown School Committee. (Women met with men as ministers and overseers on quarterly and yearly meeting levels but this is beyond the subject of this paper.) In 1877 the suggestion came from the men's Yearly Meeting that a committee of women Friends be developed to work with their brethren on the question of providing Friends' children with a religiously guarded education. In 1890, the Indian Committee came to Yearly Meeting with the request that a few women Friends be added to its ranks. In 1902 a men's Yearly Meeting committee invited a few women Friends to meet with it on the subject of marriage, and whether Friends should continue to be disowned for marrying out of Meeting. Women Friends heard with interest the report of the Meeting for Sufferings, and were especially concerned about the work of temperance, of military drill in the schools, capital punishment, and the resettlement of the Russian Doukhoborski, or Doukhobors. They did not, however, take any action on social problems as a Meeting until early in the twentieth century.

---

7. Christina H. Jones, *American Friends in World Mission,* Elgin, Ill., Brethren Publishing House, 1946.

A Widening Path

In 1908 a minute from Abington Monthly Meeting was read in the women's Yearly Meeting, suggesting that a position be taken against "underpaid" labor, a reference to the women and children in sweatshops whose condition was disturbing consciences at the time. The Yearly Meeting promptly minuted its belief that "it is not in order for this meeting to take an official action." However, a query was prepared embodying this concern, to be answered only by the "consciences of members."

A feeling that Yearly Meeting women might become active in social issues arose in 1911, as members responded to the Seventh Query: "It was asked whether, as a body of women with influence, we should not be more active in the cause of Peace, especially at this time. Let us do with our might what our hands find to do, not only in regard to peace, but to discourage intemperance, and the spirit of gambling and card playing which is so common even among women. Also to remember our duty in the concern that so largely prevails in regard to child labor." Later in the same meeting, the women suggested that a joint committee of men and women Friends be appointed to prepare a letter to send Congress in regard to the warlike attitude of this country toward Mexico; this appears to be the first initiative taken by the women's meeting since 1827. Unfortunately it came to nothing. "Fuller information in regard to the oversight of the situation by the Representative Body was presented to our meeting by men Friends, which, coupled with the changed attitude reported in the morning papers, appears to make further action unnecessary at this time."

But the winds of change were now blowing. Later in the same Yearly Meeting, after the report of Representative Meeting was read, several women suggested that the time had come to "consider the propriety of suggesting the appointment of a few women Friends as members of our Representative Body, or of appointing a small standing committee in our own Meeting to which some of the subjects claiming the deep interest and concern of many of our members might be referred for its judgement and care. Much earnest thought was given to the subject, but the way did not open for any action therein."

The question was not resolved until 1914 when the initiative came from Representative Meeting itself, suggesting that members be appointed without regard to sex. Speaking for this liberalization in the 1914 session, Joseph Elkinton insisted that "service, not sex," be the basis for representation on all committees. Immediately the education committee asked for the nomination of three women; the bar against women's participation was at last broken.

If any single woman was behind this change it was Agnes Tierney. As a young Friend she had begun to question the inequality of the status of men and women in the Religious Society of Friends as early as 1899. A strong advocate of peace, her name came forward in the Yearly Meeting in the war years, 1914–1918, as she urged women in the Yearly Meeting to play a more active role in the service of peace. In 1915 the Yearly Meeting women joined the men in action against a bill introducing military training into the schools. In 1916, women were nominated to the standing committee on peace, and in 1917 to a joint committee to consider "What part the Society of Friends can play in the present day applications of the Kingdom of God."

In 1918 a number of women, including Agnes Tierney, were appointed to work with a similar men's committee on a message on the historic peace witness of Friends. At the same meeting a representative of men's Yearly Meeting, James Moon, appeared before the women to deliver a message on women's unique position "as equal with men in our branch of the church."

While some women who were deeply concerned about peace carried the Yearly Meeting into a more active phase, others felt stirred by the problems of race relations. In 1919 Esther M. Smith brought forward a concern about the lynchings of black people in the South. She was encouraged to bring this concern to men's Yearly Meeting. Two years later, in 1921, she proposed a joint committee on race relations, which the men readily agreed to.

By 1922, men and women were serving on so many committees together in the Orthodox Yearly Meeting that holding separate sessions of yearly meeting no longer made much sense. In that year several joint sessions were tried. By 1923 the two groups met together, and separate women's meeting minutes were no longer recorded. Through 1929 a women's clerk was appointed at each Yearly Meeting, but her role became that of merely signing the epistles along with the clerk of men's Yearly Meeting. By 1930, the two had merged.

A Widening Path

It would appear, in summary, that the development of a concern for the equality of women arose in the Orthodox Yearly Meeting as a byproduct of a growing interest and commitment to social concerns, primarily the peace movement. The capacity to be sensitive to the needs of others, so much at the heart of Quaker faith and practice, proved the source of a growth impulse that overcame whatever influences had caused Orthodox women to inhibit their self-expression, and to take their rightful places at last as equal members in the Religious Society of Friends.

Hicksite Women Agitate for Equality, 1827-1922

While the Orthodox women spent the first annual meetings after the separation of 1827 lamenting the separatists and the state of the Society, the Hicksite women patterned their Yearly Meeting in 1828 exactly on its predecessors, giving no hint in the recorded minutes of any troubles at all. Their principal concern at first was establishing an exchange of epistles with the women of the other yearly meetings which had become Hicksite—first Baltimore, then Ohio, then Indiana. They also joined the men in attempting twice to establish contact with London Yearly Meeting, pointing out that they were in the majority, and were just as faithful to Christian doctrine as were the Orthodox group.

In 1830 when the second attempt was made, Lucretia Mott was in the chair as clerk of women's Yearly Meeting. Opposed to the epistle to London, which she felt was too creedal a statement for Quakers, she vacated the chair and tried to persuade the women to change the language of the epistle as prepared for the men by John Comly, the clerk of men's Yearly Meeting.[8] This unheard-of rebellion was not accepted by the women, but it marked the beginning of an era when Hicksite women, under the leadership of Mott, Deborah Wharton and several others, were willing to occasionally challenge the men's authority.

---

8. This story is told in a number of places, e.g. Anna Davis Hallowell, *James and Lucretia Mott, Life and Letters,* Boston: Houghton, Mifflin & Co., 1884, p. 107. The epistle in question in the hand of John Comly and corrected by Lucretia Mott is at the Friends Historical Library, Swarthmore College.

Friends in the Delaware Valley

In 1833 the Yearly Meeting appointed a joint committee on Education called "The Committee of Religious Concern for the Right Education of Children." Feeling, just as the Orthodox did, that too many schools were in the hands of the other branch, the Hicksites decided to make a thorough survey of all school age children within the Hicksite Yearly Meeting. This early school survey led to a recommendation that more teachers be trained to conduct Hicksite primary schools. What were needed were academies to train Hicksite teachers. In time, this concern led to the establishment of Swarthmore College and of the George School. Both men and women served on the education committee, and its reports were regularly read and discussed at women's Yearly Meeting. This was from the first one substantive issue with which that body dealt, as well as the routine of answering queries, writing epistles, and keeping accounts.

Another subject which began to be discussed at the women's Hicksite Yearly Meeting was slavery. In 1836 the women of Caln Quarter brought forward to women's Yearly Meeting a concern to send a memorial to Congress against slavery, and the application of Arkansas for admission as a slave state. A committee was appointed to consider this concern. It reported the following year that the Yearly Meeting should not act as a body on such issues, but that antislavery action should be left to individual conscience. Nevertheless, the subject was brought up again and again, not banned from discussion as at the Orthodox meetings.

A third subject much discussed in the Hicksite women's Yearly Meeting in the early days of its existence was the situation of the American Indians. Both men and women served on the Yearly Meeting Indian Committee, which with the New York and Genesee Hicksite Meetings established a center and a school for the Senecas at Cattaraugus. In 1838 the women's meeting also considered joining with the men in protesting to Congress their treatment of the Cherokees in Florida. These and similar issues were debated in the years that followed.

In 1841 Caln Quarter, often the innovator, brought to the women's Yearly Meeting the proposal that the rules of discipline be so altered as to give women more authority in the disciplining of their own members. This proposal reflected a concern first voiced by Lucretia

A Widening Path

Mott, that it was wrong to ask women elders to "deal" with offenders and then give the final disposition of the case to the authority of the male elders. The women proposed a number of their members to serve with the men on a committee to examine the proposed change, but word was brought back from the men's Yearly Meeting that "the way did not open" for the matter to be pursued.

This appears to be the first expression of a concern for equality within the women's Hicksite Yearly Meeting. In 1844 the Yearly Meeting's committee on Education minuted that it had received reports from all the schools under the care of the Yearly Meeting which were administered by joint committees of men and women. "Would not advantages accrue if such joint committees were more general?" the women queried.

Continuing with their tradition for independent action, the women's Yearly Meeting in 1855 proposed an address to the southern states in regard to slavery. The men's Meeting responded that way did not open. However, the following year they apparently reversed themselves, for such an epistle was prepared.

Since shortly after the separation, the Hicksites in Philadelphia had been holding Yearly Meeting in two separate meeting houses; the men at Sixth and Green, and the women at Fifth and Cherry. Though the women were slightly in the majority, this meeting house was smaller, noisier, and harder to ventilate. In 1855 the women wrote a strong minute of protest:

> The Yearly Meeting of Women Friends, now sitting, believe it right to call the attention of men Friends to the very poor accommodations the Cherry Street house affords. Although the additional ventilation has rendered it a little more comfortable, still its crowded state, the difficulty of hearing, and our position since the erection of the adjoining building, being by many considered unsafe, we feel it right to present the subject before you.

Action resulted a month later, when the committee appointed by the men's Meeting recommended the purchase of a lot at Fifteenth and Cherry which could be acquired at an advantageous price. By fall, fundraising was underway. Begun in 1856, the building was ready for occupancy by the time of the May Yearly Meeting in 1857, the men's meeting informing the women that they might use the Race Street room if they would kindly houseclean it!

Emboldened by this rapid success perhaps, the Hicksite women

Friends in the Delaware Valley

took the initiative again a few years later, proposing to the men's Yearly Meeting a change in discipline in regard to marriage which would permit the person proposing marriage with an attender to lay the matter before Friends. (Previously such marriages were grounds for disownment.) The men responded that way did not open. This answer did not deter the women, who raised the question in 1861 and were again rebuffed. At the same time some changes in the discipline governing women's meetings on all levels were agreed to, giving women more authority.

The Civil War was a time of testing for the Hicksites, many of whom had supported the abolition of slavery and found their sympathies with the Northern cause. Some younger Hicksites chose to fight, and were disowned. The women's Yearly Meeting minutes during this period reflect some of this tension. War was deplored, as well as capital punishment, and conscientious objection was supported, but each forward step toward emancipation of the slaves was greeted with pleasure. Outside of Yearly Meeting, the women began packing boxes for the contrabands, just as their Orthodox sisters were doing, and carried the burden of fundraising for this enterprise from April of 1862 until January of 1864, when the Association for the Aid and Elevation of Freedmen was formed. Both men and women belonged to this association, which supported Cornelia Hancock, a young New Jersey Quaker from Salem Quarter, who went south after Gettysburg and established a school in Mt. Pleasant, South Carolina, across the bay from Charleston; and Martha Schofield, of Bucks Quarter, who worked in several locations and finally established a school for blacks in Aiken, South Carolina.

The war years and the concern for reconstruction did not deter the Hicksite women from their drive toward equality. In 1867 Concord Quarter brought a proposal to women's Yearly Meeting that the discipline be changed to give women equality. Lucretia Mott was assigned to a committee to consider the change. A message from men's Yearly Meeting, saying that way did not open, was signed by James Mott!

The women's campaign continued. In 1869 in the summary of various spiritual messages delivered at women's Yearly Meeting it was said that "the growing acknowledgement of the true position of women was frequently alluded to, and we were exhorted not to allow our domestic duties to absorb all our attention." In 1870, in answer

A Widening Path

to the Fourth and Fifth Query, the women noted that, "It was believed that the influence which a woman can exert when led by the Holy Spirit was not sufficiently estimated." At this meeting the proposal that women be chosen for Representative Meeting on an equal basis with men was first raised. The next year the women appointed a committee to discuss the proposal with the men, who had agreed to this action. In 1872 the women criticized Representative Meeting for not "remonstrating against the practice of settling difficulties, national and otherwise, by resorting to arms." In a formal minute the women urged joint representation of men and women on the all important Representative Committee. "Male and female being one in Christ, the spiritual equality of men and women being fully recognized by our Society, there would seem to be a propriety in the care of the Church devolving on the whole body."

In 1874 the men's Meeting responded to this growing agitation with the message that they saw no need for a change in discipline; there was no reason now why the names of suitable Friends of either sex might not be brought forward. This was still not full equality; in 1875 Radnor Monthly Meeting suggested that the discipline be changed so that "Women Friends would have the same voice as men in all business meetings of the Society." A committee of men and women was appointed to consider this change, and finally, in 1877, the discipline was changed accordingly and women were appointed in equal numbers to Representative Meeting.

The early recommendation of the Yearly Meeting Committee on Education that more teachers be trained to staff the Hicksite schools had led to the establishment of a committee of Friends from both Baltimore and Philadelphia Yearly Meetings to consider the need for a boarding school or college. This committee, it was decided, would become a body independent of the Yearly Meetings. Both James and Lucretia Mott served on the committee, which began meeting in Baltimore in 1860 and led to the chartering of Swarthmore in 1864, and its opening as a coeducational academy and college in 1869. Friends were pleased with this development, but still desired a boarding school of their own under the care of the Philadelphia Hicksite Yearly Meeting. In 1879 the women decided to ask the men's Yearly Meeting to unite with them in appointing a committee to plan such a school.

Lucretia Mott, now eighty-six, was in attendance at this meeting, and it was probably she who gave an anonymous message, urging women to live up to their new responsibilities: "The position of women of this Society now admitted to full participation in the active engagements of the church should be appreciated by us. We should prove by our efficient service to the cause of justice, mercy and truth that we are indeed in our right place, co-workers with our brethren and earnest promoters of reasonable and truly religious progress."

The concern for the establishment of a coeducational boarding school, thus begun in 1879, was expressed regularly for almost a decade. In 1888 it was announced that the estate of John George would make the purchase of land feasible. The George School opened its doors in 1891. The committee which administered it was of course made up of men and women on an equal basis.

In addition to these concerns for their own equality, and for equal educational opportunities, the Hicksite women continued to manifest the breadth of their social concerns. The reports of the Indian Committee were discussed, and support for schools for the newly freed slaves was urged. The developing concern for temperance found its way into the women's minutes several times. In 1877 the women decided to address Lucy Hayes, First Lady, and thank her for not serving wine at White House functions. In 1892 the women proposed a remonstrance to the World's Fair committee in Chicago against the sale of liquor.

The Peace Testimony also absorbed the interest of the Hicksite women's Yearly Meeting, with testimonies against war expressed in 1860, 1870, and 1872. In 1882 the women protested the military display in connection with the Bicentennial celebration of the landing of William Penn. In 1894, they urged a memorial against military training in the schools, resulting in a minute from the whole Yearly Meeting addressed to the Philadelphia Board of Education in 1897. In 1899 the women expressed their hopes for the peace conference meeting in The Hague.

By the 1890's, a Committee on Philanthropic Labor had been created to deal with issues of "social purity" and "prejudice against the colored." The women concerned themselves with protesting the growing of tobacco, and tried in vain to persuade men's Yearly Meeting to join with them in a protest against the polygamy of the Mor-

A Widening Path

mons. They also spoke out against the regulation of prostitution, or "vice," as they called it, fearing that this gave it legitimacy.

In 1893 Genesee Yearly Meeting suggested that the men and women of Philadelphia Yearly Meeting (Hicksite) might consider meeting as one. The Hicksite women, however, were satisfied with things as they were, although in 1896 they took one more step in urging that women be appointed as Meeting trustees. With many social concerns to occupy their energies, they entered the Twentieth Century by reaffirming their peace testimony and preparing to memorialize Congress on the subject of the sale of liquor. In 1909 they were appointing Hannah Clothier Hull to represent the entire Yearly Meeting at a Peace conference. The next year they finally persuaded the men to join them in a memorial to Congress against the white slave traffic.

The issue of women's suffrage began to appear in the minutes of the women's Yearly Meeting in 1914. In 1919 they minuted the connection they saw between intemperance and inequality: "that large sums of money are raised by the liquor interests to combat suffrage for women is proof that they fear the women's vote."

By this time the women were regularly hearing reports from the Committee on Philanthropic Labor, which was turning its attention to an anti-lynching crusade; a Peace Committee, the Central Bureau (a group formed to coordinate all the committees of Race Street Yearly Meeting) and the newly formed American Friends Service Committee. It was also at last beginning to communicate with London. Many Race Street Friends were involved in Friends General Conference, and there was beginning to be more communication with the Orthodox Friends at Arch Street. The women's Yearly Meetings were becoming more lively, although they were still conducted with great solemnity.

In 1911, an unusual woman had been hired to staff the Central Bureau of Race Street Yearly Meeting. She was Jane Rushmore of upstate New York, a teacher and social worker. Jane was chosen as recording clerk of the Yearly Meeting for Women in 1918 and continued in this position until 1922, when she was made clerk.[9]

In 1921 a minute had come in from the Western Quarter proposing

---

9. Emily Cooper Johnson, *Under Quaker Appointment, The Life of Jane P. Rushmore.* Philadelphia: University of Pennsylvania Press, 1953.

once again that men and women meet jointly. The Race Street Friends were dubious about the change, but they decided to try one joint session. In 1922, Jane Rushmore's first year as clerk of women's Yearly Meeting, joint sessions were held on two days, and in 1923, on three. Then in 1924 the two groups met separately long enough for each to approve adjourning to meet as a single body. The representatives at this joint meeting then retired, and came back with the unheard of recommendation that a woman be the clerk for the combined Yearly Meeting. Jane Rushmore thus became the first woman to fill such a position in Philadelphia Yearly Meeting history. She remained clerk for two more years, and continued to be a strong influence for many years after that in Yearly Meeting.

Summary

Several interesting themes appear to emerge as we review this two hundred thirty-odd years of the history of Philadelphia Yearly Meeting of Women. One is that the women had from the first a strong sense of function; the keeping of accounts and minutes, the selection of officers, the exchange of epistles. They also felt keenly about their place of meeting and were willing to be insistent on an adequate meeting space at a time when they generally gave way to the authority of the men's Meeting.

Another matter on which women felt strongly enough to take initiative was always the welfare of their sisters. The campaign of Philadelphia Yearly Meeting women for British women Friends to have their own yearly meeting is quite striking in contrast with the self-restraint with which they approached other issues at that time. The sending of "stock" to aid suffering sisters, and the general interest in the exchange of epistles with other women's meetings, also suggested the strength of this feeling. In the Hicksite minutes particularly, it is to be noted that a concern for Indian women, slave women, women who were victims of white slavery or polygamy, was a motive force in causing the Quaker women to seek social action.

The care of children, and the conduct of schools was an area in which women felt themselves to have a vital stake, and in which they sought some voice. During the long period when the Orthodox

A Widening Path

women felt themselves under religious and social constraint, their only form of expression was participation in the Westtown Committee and an interest in the school in Tunesassa. Among the Hicksite women, the lively Committee on Education seems to have been a training ground for them to become effective on other joint committees.

Finally, it seems clear that it was social concern—for foreign missions, child labor, and for peace among the Orthodox women; for the abolition of slavery, temperance, race relations and peace among the Hicksite women—that led them eventually to see the need to play a larger role in the life of the church. To work against slavery, to protest military training in the schools, to object to child labor, women were willing to assert their own rights.

In 1839, Abby Kelley, a Massachusetts Quaker, wrote to Lucretia Mott to question the validity of the American Convention of Anti-Slavery Women, which seemed to her separatist. In response, Lucretia defended the separate meetings for women within the Society of Friends: "And yet their (Friends) meetings for women, imperfect as they are, have had their use, in bringing our sex forward, exercising their talents and preparing them for united action with men, as soon as we can convince them that this is both our right and our duty."[10]

If men and women had met together from the first, in the business meetings of the Society of Friends, it is probable that women would not have spoken up, as both Fox and Penn feared. They would certainly not have learned to conduct business, keep minutes, exchange epistles, keep accurate accounts. All this training led to the important roles individual Quaker women played in the reform movements, and pioneering in the professions during the nineteenth century. Many historians now credit the emergence of the early women's rights movement to the leadership of Quaker women. Today, at last, the Society of Friends is struggling to understand and implement the full implications of the ancient Quaker belief in equality, based on the biblical assertion that in Christ there is neither male nor female.

---

10. Lucretia Mott to Abby Kelley (Foster), Mar. 18, 1839. Abby Kelley Foster papers, American Antiquarian Society, Worcester, Mass.

VI-A Philadelphia Friends and the Indians

by MILTON REAM

George Fox set a good example for Friends by taking every opportunity to visit the Indians and to preach to them. He put Quaker and Indian relationships on a sound footing by proclaiming that all men could respond to God. In 1682 Fox wrote "An epistle to all planters, and such who are transporting themselves into foreign plantations in America, &c." He urged those making "outward plantations" to "invite all the Indians, and their kings, and have meetings with them, or they with you; so that you may make inward plantations with the light and power of God, (the gospel) and the grace, and truth, and spirit of Christ; and with it you may answer the light, and truth, and spirit of God, in the Indians... that they may serve and worship him, and spread his truth abroad."[1] In 1685 Burlington Quarterly Meeting appointed a committee to pay a religious visit to the Indians. Their appointed meeting was well attended by quiet and attentive Indians.[2] In 1687 Fox wrote "To Friends in West Jersey and Pennsylvania" urging them to have meetings with the Indians, "to let them know the principles of truth; so that they may know the way of salva-

---

1. *A Collection of Many Select and Christian Epistles, Letters and Testimonies, Written on Sundry Occasions, by that Ancient, Eminent, Faithful Friend, and Minister of Christ Jesus, George Fox* (Philadelphia: Marcus T.C. Gould, 1831), v. II, p. 218.
2. John Gough, *A History of the People Called Quakers* (Dublin: Printed for Robert Jackson, 1789), v. 3, p. 308.

tion, and the nature of true Christianity, . . . and how that Christ hath enlightened them, who enlightens all that come into the world. And God hath poured out his spirit upon all flesh; and so the Indians must receive God's spirit; . . . And so let them know, that they have a day of salvation, grace, and favour of God offered unto them; if they will receive it, it will be their blessing."[3]

William Penn shared George Fox's views as to the spiritual kinship between himself and the Indians. In his 1681 letter to the Pennsylvania Indians, he reminded them that they were created by the same God who "hath written his law in our hearts, by which we are taught and commanded to love and help and do good to one another." He continued that he had been given a great province, "but I desire to enjoy it with your love and consent, that we may always live together as neighbours and friends."[4] Few colonizers shared Penn's hope that Indians and whites would live as neighbors in the same community. A century and a half later, it was generally assumed that the place for Indians was beyond the frontier or on a reservation.

In planning the government of his colony, Penn laid the foundation for peaceful coexistence by proposing: (1) Before settlements were made on any tract of land, commissioners should clear title to the Indians' satisfaction; (2) the justice system would be open to both races to discourage private revenge. Juries hearing inter-racial cases would be composed of six men from each race; (3) the government would regulate trade with the Indians to prevent them from being cheated; (4) no person could sell any strong liquor to the Indians.[5] The respect with which Penn treated the Indians made a deep impression and, for years afterwards, Quakers enjoyed the benefits of his reputation.

Evidently the colonial government had difficulty enforcing William Penn's proposals regarding Indian–white relationships because Philadelphia Yearly Meeting began to make them matters of discipline. The Yearly Meeting's first advice concerning Indians was issued in 1685 and stated "that it is not consistent with the honour of Truth for any that makes profession thereof to sell rum or other

---
3. Fox, *Op. cit.*, v. II, pp. 305-306.
4. Albert Cook Myers, *William Penn: his own Account of the Lenni Lenape or Delaware Indians, 1683* (Moylan, Pa., 1937), p. 66.
5. *Ibid.*, pp. 58-59, 66-67, 74, 85.

Friends in the Delaware Valley

strong liquors to the Indians, because they use them not to moderation, but to excess and drunkenness." This advice was repeated the following year and in 1687 the Yearly Meeting went further in requiring, "that this our testimony may be entered in every monthly meeting book, and every friend belonging to the said meeting to subscribe the same."[6] Surviving statistics from three monthly meetings record that seventy-seven male members signed at Chester, seventeen at Concord, and forty-nine at Middletown. The Yearly Meeting found it necessary to warn Friends again regarding this matter in 1719 and 1722.[7]

A second concern regarding Indians came from the Yearly Meeting of 1719. While repeating its earlier advice regarding the sale of liquor to the Indians, the Yearly Meeting added, "to avoid giving them (Indians) occasion of discontent, it is desired, that Friends do not buy or sell Indian slaves."[8]

The Yearly Meeting of 1763 took up the matter of land purchases from the Indians. "Friends should not purchase, nor remove to settle in such lands, as have not been fairly and openly first purchased from the Indians, by those who are, or may be, fully authorized by Government to make such purchases." Monthly Meetings were to insist on "the strict observance of this advice" and if Friends still prepared to occupy such lands, the monthly meeting was to withhold certificates of removal.[9]

The printed Discipline in 1797 included the 1722 advice regarding selling liquor to the Indians and the 1763 advice warning against settlement on lands not properly purchased from the Indians. It also included a 1759 reminder of God's blessings on the first settlers who "were comforted in times of want and distress" by the Indians.[10] The

---

6. A Collection of Christian and Brotherly Advices Given Forth, From Time to Time, by the Yearly Meetings of Friends for Pennsylvania & New-Jersey held alternately at Burlington and Philadelphia, Alphabetically Digested under Proper Heads, 1762 [manuscript] The Quaker Collection, Haverford College. pp. 79–80.
7. Joshua L. Baily, "Progress of the Temperance Cause among Friends of Philadelphia," *Bulletin of the Friends Historical Society of Philadelphia,* v. 1, no. 1 (Tenth month, 1906), p. 25.
8. A Collection of Christian and Brotherly Advices, 1762, p. 79.
9. A Collection of Christian and Brotherly Advices given forth from time to time by the Yearly Meeting of Friends for Pensylvania [sic] and New-Jersey held alternately at Burlington & Philadelphia, alphabetically digested under proper heads, n.p., n.d. [manuscript] Quaker Collection, Haverford College, Haverford, Pa.
10. *Rules of Discipline and Christian Advices of the Yearly Meeting of Friends for Pennsylvania and New Jersey* (Philadelphia: Samuel Sansom, 1797), pp. 60–61.

Philadelphia Friends and the Indians

printed Discipline of 1806 omitted advices relating to Indians, possibly reflecting their disappearance from the Delaware Valley.

Westward expansion and the French and Indian War eventually caused a major conflict between the colonists and the Indians. In 1756 the Governor and Council of Pennsylvania declared war on the Indians. Israel Pemberton and other Friends, concerned that Quaker methods of dealing with the Indians were no longer the policy of the government, founded the Friendly Association for Regaining and Preserving Peace with the Indians by Pacific Measures. A request for funds produced more than £3,000 from Quakers and another £1,500 from Mennonites and Schwenkfelders.

At first the main function of the Friendly Association was to attend gatherings of Indians and government officials where Friends sought to reassure the Indians by their presence that no undue tactics would be used and that all their grievances would be aired. Quaker delegations attended two conferences between the Governor and the Delawares in 1756 and a third conference in 1757. The Friends gave presents to the Indians, entertained and advised them, and made transcripts of the proceedings.

A major issue at these conferences was the Walking Purchase. The purchase was alleged to have been negotiated in 1686 in a deed allowing the Proprietor to acquire land for the distance that a man could walk in a day and a half. The Indians were to occupy the land until such time as the Proprietor needed it. In 1737 the Proprietor, Thomas Penn, claimed the land and more than doubled the area by hiring trained young men to "walk" over a cleared path. The enlarged area included the homeland of some Delawares at the fork of the Delaware and Lehigh Rivers. They refused to move until forced to do so by their Iroquois overlords. Because the Friendly Association backed the Delawares' claims of fraudulent land sales, the Proprietary party did not welcome their presence at treaty negotiations.

A fourth conference in 1758 was attended by more tribes. At this larger conference and in the presence of their traditional overlords, the Iroquois, the Delawares had less influence. In 1762 the tribe accepted payment for their land claims and in return dropped their charge of fraud against the Proprietors. The withdrawal of the charge was regarded as a major defeat for the Friendly Association which had backed the Indians' claim.

Friends in the Delaware Valley

Other Association activities included support of Indian requests for a Delaware reservation in the Wyoming Valley. In 1758 the Friendly Association loaned the Indians money to complete cabins on the reserve. At a treaty with Ohio Indians at Lancaster in 1762, they provided money and goods to ransom captives. They backed a system of government-operated trading posts, but hesitated to invest any of their funds. The Association sponsored a Moravian teacher, Christian Frederick Post, who spent a short time with Indians near Ft. Pitt.

The Friendly Association probably faded away about 1764. It had not found a basis for permanent peace but it had served as an expression of Quaker good will toward the Indians when that attitude was no longer the policy of the provincial government.[11]

New Jersey Friends established in 1757 an organization similar to the Friendly Association known as the New Jersey Association for Helping the Indians. The group hoped to buy a tract of land for the use of landless Indians. In 1758 the government of New Jersey agreed to pay the Indians for their land claims and the Indians used the money to buy 3,000 acres. Since its purpose had been accomplished by other means, the Association for Helping the Indians was laid down.[12]

Quaker encounters with the Indians became fewer as they moved west ahead of expanding white settlements. A few Friends travelled long distances to maintain symbolic ties. In 1773 Zebulon Heston and John Parish travelled 120 miles west of the Ohio River, 450 miles from Philadelphia, to visit the Indians. The Meeting for Sufferings of Philadelphia Yearly Meeting took the opportunity to send a message expressing their wish to maintain "the old friendship which was made between your fathers and ours" and urging the Indians "to

---

11. Samuel Parrish, *Some Chapters in the History of the Friendly Association for Regaining and Preserving Peace with the Indians by Pacific Measures* (Philadelphia: Friends' Historical Association of Philadelphia, 1877).

    Theodore Thayer, "The Friendly Association," *The Pennsylvania Magazine of History and Biography,* October, 1943, pp. 356-376.

    Rayner Wickersham Kelsey, *Friends and the Indians, 1655-1917* (Philadelphia: The Associated Committee of Friends on Indian Affairs, 1917), pp. 49-51.

    Sydney V. James, *A People Among Peoples: Quaker Benevolence in Eighteenth-Century America* (Cambridge: Harvard University Press, 1963), pp. 178-192.

12. Kelsey, pp. 46-47.

    James, pp. 200-201.

attend diligently to the instructions of the Spirit of Christ within you." As for the Indians' request to have teachers sent to them, Friends approved the idea and "whenever we can find any rightly qualified and willing to undertake the service, we intend to assist and encourage them in it." More than twenty years would pass before Friends sent teachers. In 1791 the Seneca chief, Cornplanter, requested teachers and in the meantime sent three boys to be educated by Friends in Philadelphia. In 1793 the Meeting for Sufferings sent representatives to the treaty negotiations at Sandusky, Ohio. That same year two Friends visited the Delawares near Muskingham. In 1794 four Friends represented the Meeting for Sufferings at treaty negotiations with the Six Nations at Canandaigua, New York. Chief Sagareesa repeated the Senecas' request for teachers. This request was rather vaguely passed on to the Meeting for Sufferings in the report of the four Friends who believed, "some mode may be fallen upon of rendering them more essential service than has yet been adopted."[13]

The next year, 1795, the Indian concern was brought before the Yearly Meeting which recorded that "our minds have been measurably drawn into sympathy with these distressed inhabitants of the wilderness, and on comparing their situation with our own, and calling to grateful remembrance the kindness of their predecessors to ours in the early settlement of this country, considering also our professed principles of peace and good will to men, we were induced with much unanimity to believe, that there are loud calls for our benevolence and charitable exertions to promote amongst them the principles of the Christian religion, as well as to turn their attention to school learning, agriculture, and useful mechanic employments especially as there appears in some of the tribes a willingness to unite in the exercise of endeavours of this kind."[14] Where earlier interest in Indians had found expression in individual concerns, in unofficial organizations like the Friendly Association, and in occasional

---

13. *Some account of the Conduct of the Religious Society of Friends towards the Indian Tribes in the Settlement of the Colonies of East and West Jersey and Pennsylvania: with a brief narrative of their labours for the Civilization and Christian instruction of the Indians, from the time of their settlement in America, to the year 1843.* Published by the Aborigines' Committee of the Meeting for Sufferings (London: Edward Marsh, 1844), pp. 94-113.
14. Minutes of Philadelphia Yearly Meeting, 1780-1798, p. 317 [Microfilm in the Quaker Collection, Haverford College.]

actions by the Meeting for Sufferings, Philadelphia Yearly Meeting in 1795 felt led to appoint a standing Indian committee, the Committee for Promoting the Improvement and Gradual Civilization of the Indian Natives. Over nearly two centuries the service of the Indian Committee has taken a variety of forms, some sanctioned by long practice and others requiring new insights and commitments.

The committee soon entered into a new form of service, Indian education. The committee canvassed the Indians of Pennsylvania and New York to see if any were receptive to Quaker service among them; the Oneidas of New York were. With the approval of the government, the committee sent three Friends in 1796 to teach literacy, agriculture, and the principles of Christianity. The Friends held reading classes, repaired a ruined sawmill, and distributed farm tools. Most of their effort went into their model farm. Two women Friends came to teach spinning, knitting, sewing, and housekeeping. A blacksmith taught his trade while keeping the farm tools in repair. After a while some Indians began to say the Quakers were interested in acquiring land, so, in 1799 the mission was closed leaving the farm and its equipment to the Oneidas.[15]

In 1798 the Indian Committee opened work among the Senecas when Joel Swaine, Henry Simmons, Jr., and Halliday Jackson moved to Cornplanter's village on the Allegheny River, near the Pennsylvania-New York border. This mission, like that to the Oneida, emphasized the model farm and progress toward subsistence farming.

The work of the new mission was soon disrupted by the religious visions of Handsome Lake whose teachings divided the tribe into antagonistic groups. When the Indians asked Halliday Jackson's opinion of the revelations, he decided that the new religion called for useful reforms of behavior and replied diplomatically that he thought that they "would do well to observe the sayings."[16]

In 1803 the mission was moved to land purchased by the Committee near the reservation. A site was selected a few miles to the north where Tunesassa Creek flowed into the Allegheny River. A sawmill

---

15. James, pp. 303-304.
   Some account . . ., pp. 115-118.
16. James, pp. 304-307.

Philadelphia Friends and the Indians

and a gristmill were built and the model farm was reestablished. Women were instructed in housekeeping, spinning, weaving, and knitting. Although there had been earlier educational efforts, Joseph Elkinton is credited with starting the school for Indian boys in 1822. In 1825 Mary Nutt established a school for girls. At times additional schools were maintained in nearby villages. In 1852 the boarding school was established which was finally discontinued about 1939. The facilities continued to be used for vocational training, homemaking and sewing classes, a health clinic, and as a meeting place for community groups.[17]

For a time Philadelphia Friends maintained additional work in New York. In 1799 Senecas living forty miles northwest of Tunesassa, on the Cattaraugus River, asked the Philadelphia Yearly Meeting Indian Committee to help them set up a sawmill. This was done. In 1803 the Friends who selected the site for the Tunesassa mission visited the Senecas in the Cattaraugus area. In 1808 a farm was purchased at Clear Creek and in 1809 four Friends arrived to teach agriculture to the Cattaraugus Indians. The work followed the earlier pattern and included saw and gristmills and instruction in farming and housekeeping. The mission at Cattaraugus was laid down in 1815 but one of the Friends, Jacob Taylor, bought part of the farm and continued to live there until 1821.[18]

Later work at Cattaraugus was under the care of Hicksite Friends. Philadelphia Yearly Meeting (Hicksite) joined with Genesee, New York, and Baltimore Yearly Meetings (Hicksite) to operate a Female Manual Labor School among the Senecas at Cattaraugus. The school, in operation from 1847 to 1849, taught housekeeping skills to Indian girls.[19]

A new form of service opened for Friends when President Grant as a part of his Peace Policy asked Friends to administer two Superintendencies within the Bureau of Indian Affairs. The Peace Policy has been defined as, "Grant's radical innovations of 1869 and 1870: the appointment of Ely S. Parker, church control over agents, creation

---

17. Kelsey, pp. 97–104.

    Proceedings of the Yearly Meeting of the Religious Society of Friends of Philadelphia and Vicinity, 1939, pp. 120–124. (Orthodox)
18. Kelsey, pp. 104–105.
19. *Ibid.,* pp. 125–130.

of the Board of Indian Commissioners, and a greatly intensified program of federal aid to Indian education and missions."[20] Orthodox Friends were offered the Central Superintendency which consisted of the tribes in Kansas and the Indian Territory (Oklahoma) except for the Five Civilized Tribes. Hicksite Friends were assigned the Northern Superintendency composed of the six agencies in Nebraska.

Friends saw an opportunity through a fair, humane, and kindly Indian administration to bring the Indians "to willingness to adopt the arts of civilized life, and listen without prejudice, to the truths of Christianity."[21] They hailed the opportunity for consistent Friends "not only to improve the material condition of the Indians, but to exemplify to the world the wisdom and profit of acting in strict accordance with the benign principles of the gospel of salvation, in all transactions between man and man, and equally in the intercourse between nations and tribes."[22]

As an indication of specific goals Friends hoped to accomplish, the outline of measures "to promote civilization" drawn up by the Hicksite Friends could serve equally well for the Orthodox administration:

> 1. Care to recommend for appointment . . . such persons only as seemed to be properly qualified for the position, and whose moral influence would promote the growth of virtue. . . . Each agency employee was appointed for practical missionary work and expected to be a missionary for good in precept and example.
> 2. The establishment of schools . . . taught by Christian teachers. Sabbath schools were held at all the agencies, in which Scripture lessons blended with religious instruction were given. . . .
> 3. The allotment of lands in severalty. . . .
> 4. The instruction of Indians in agriculture, . . . in mechanical employments and in household pursuits.
> 5. The distribution of agricultural implements, live stock and seeds.
> 6. The building of dwelling houses, and planting of fruit trees. . . .
> 7. The employment of matrons to instruct the Indian women in household duties and the care of the sick.

---

20. Robert H. Keller, The Protestant Churches and Grant's Peace Policy: a Study in Church–State Relations, 1869–1882. Dissertation. Divinity School of the University of Chicago, 1967. p. 69.
21. *The Friend* (Philadelphia), v. 42, no. 33 (4/3/1869), p. 256.
22. *Ibid.*, p. 255.

8. . . . to speedily advance the condition of the Indians to the status of Christian, educated, self-supporting American citizens, living in comfortable houses on lands held by them in fee simple.[23]

Friends soon organized to assume their new responsibilities. Orthodox Friends met at Damascus, Ohio, in June, 1869, and formed the Associated Executive Committee of Friends on Indian Affairs, representing Philadelphia, New England, New York, Baltimore, Ohio, Indiana, Western Iowa, and North Carolina Yearly Meetings. Enoch Hoag from Iowa Yearly Meeting was appointed Superintendent of the Central Superintendency and nine Friends served under him as agents. Other Friends were hired as farmers, teachers, blacksmiths, and millers. Dr. William Nicholson of North Carolina was appointed general agent for the Associated Executive Committee and served as their representative in the field. The Philadelphia Indian Aid Association was formed to support the work of the Associated Executive Committee.

Except for an interval of two or three years Philadelphia Friends chaired the Associated Executive Committee for ninety years beginning with John B. Garrett, selected as the first chairman in 1869, and extending to Lawrence E. Lindley who served until 1959.[24]

President Grant made Hicksite Friends the administrators of the Northern Superintendency. Philadelphia Yearly Meeting (Hicksite) sent delegates to the general conferences of the seven yearly meetings overseeing the work. When Superintendent Samuel M. Janney retired in 1871, Barclay White from Philadelphia Yearly Meeting was appointed in his place. He held the office until 1876 when the Northern Superintendency was abolished and his agents reported directly to the Commissioner of Indian Affairs.

President Hayes was not supportive of Friends in official government positions, so in May 1879, the Associated Executive Committee resigned any further responsibility for Indian administration. Of the Hicksite agents, only two were still serving in 1880. The last Friend resigned in 1885. Hicksite Friends remained interested in the Indians

---

23. Barclay White, *The Friends and the Indians* (Oxford, Pa: Published for the Convention, 1886), pp. 9–10.
24. Kelsey, pp. 162–187.

of Nebraska and for a time around 1890 supported field matrons who taught housekeeping in the Indian homes.[25]

Friends' time as Indian administrators was too short to accomplish their goal of making yeoman farmers of the Indians. Many critics at that time and since have regarded the peace policy as a complete failure. One historian wrote, "Capability and strength determined the course of Indian-white relations. Construction of railroads, sod houses, repeating rifles, and windmills were deeds more powerful than Quaker memorials." The same historian concedes that "measured by certain standards of the period, it was a success. According to the Board of Indian Commissioners, almost one-half of all Indians wore the white man's clothing in 1876; between 1868 and 1876, houses on reservations increased from 7,500 to 56,000; schools and teachers tripled; acres under cultivation increased six-fold; and Indians owned fifteen times more livestock."[26] As Friends left office, they could note considerable material improvement among the Indians under their care. They could look back on duty faithfully and honorably discharged. They could also point to many achievements including the construction of buildings, the opening of farms on the virgin prairie, and the founding of schools, some of which still exist as government schools in Oklahoma.

The Associated Executive Committee of Friends on Indian Affairs remained in existence to oversee the educational and religious work they had started at many of the agencies. Four Friends Indian Centers remain today carrying on religious work and programs of community service including youth camps and scouting, weaving, community social activities, and prison visitation.

While engaging in new activities, the Indian Committee did not neglect old forms of service such as entertaining visiting Indians, lobbying in state capitals and in Washington on behalf of Indian interests, monitoring treaty negotiations, and intervening in emergency situations. To mention two examples from many, Philadelphia Yearly Meeting sent a memorial to Congress in 1830 on behalf of the Cherokees who were being driven from their homeland in Georgia.[27]

---

25. *Ibid.,* pp. 185-198.
26. Keller, pp. 230, 328.
27. *Some account . . .,* p. 141.

Similarly in the late 1950's the Indian Committee became concerned for the New York Senecas whose reservation would be partially inundated by the federal government's Kinzua Dam. This concern preoccupied the Indian Committee for some years.

The Indian Committee of Philadelphia Yearly Meeting (Hicksite) cooperated with those of Genesee, New York, and Baltimore Yearly Meetings (Hicksite) between 1838 and 1842 to prevent the Senecas from being defrauded of their New York lands by the Ogden Land Company.[28]

Another long established form of service undertaken by the Indian Committee was providing assistance to other committees and organizations whose aims were similar. A few examples may be mentioned: About 1805 the Philadelphia Indian Committee contributed £4,760 to the Baltimore Yearly Meeting's Indian Committee to start a mission near Ft. Wayne, Indiana. They also contributed £2,250 to the Indian Committee of New York Yearly Meeting to establish a mission at Brothertown in 1806.[29] Friends of Philadelphia Yearly Meeting have contributed their time and abilities to a number of organizations working in behalf of American Indians.

One such organization is the Indian Rights Association which was founded at Philadelphia in 1882 by Herbert Welsh and Henry S. Pancoast. The Association's purpose is "to secure to the Indians of the United States the political and civil rights already guaranteed to them by treaty and statutes of the United States, and such as their civilization and circumstances may justify." The means to this end are "influencing public opinion and Congressional legislation and by assisting the executive officers of government in the improvement of the laws passed for the protection and education of the Indians." The organization reported its on-the-spot investigations in pamphlets and maintained a lobbyist in Washington.[30]

Some Friends served on the Board of Indian Commissioners which was established by Congress in 1869 as a non-partisan organization to oversee the administration of Indian affairs. The Board was com-

---
28. Kelsey, pp. 118–125.
29. James, pp. 309–311.
   Some account . . ., p. 128.
30. *The Friend* (Philadelphia), v. 56, no. 20 (12/23/1882), p. 159.
   William T. Hagan, *American Indians* (Chicago: University of Chicago Press, 1979), p. 124.

posed of "men eminent for their intelligence and philanthropy, to serve without pecuniary compensation. . . ." They had authority only to audit, inspect, and recommend. The Board was abolished in 1933.[31]

The service of one Friend, Albert K. Smiley, on the Board of Indian Commissioners led him to found the Lake Mohonk Conference on the Indian by inviting outstanding persons concerned about Indian policy to meet annually at his resort hotel at Lake Mohonk, New York. From 1883 to 1916 the conferences "served as a clearinghouse for the ideas and strategy for reform of government Indian policy." Most of the Mohonk participants believed in individualism and work and pursued a goal of Indian assimilation through "Citizenship, Civilization, and Christianity." At the same time they exhibited "a depth of compassion, a drive for equality of opportunity, an appreciation for individual identity and cultural awareness that was far ahead of its time."[32]

A Friend who worked in an official government position was Charles James Rhoads, a member of Philadelphia Yearly Meeting, who served as Commissioner of Indian Affairs under President Hoover. Rhoads accepted the position when Hoover agreed to back the reforms suggested in the Brookings Institution's 1928 report, *The Problem of Indian Administration,* often called the Meriam Report after the editor, Lewis Meriam. Rhoads appointed another Quaker, Joseph Henry Scattergood, as assistant commissioner and the two worked closely together. Reformers like John Collier, executive secretary of the American Indian Defense Association and critic of the Indian Bureau, welcomed the Rhoads-Scattergood administration. Rhoads and Collier jointly drafted reform legislation and then disagreed as to the method of presenting it to Congress. Rhoads submitted the proposals and waited for Congress to act while Collier wanted to lobby in behalf of the reforms. The two men also disagreed over land policy. Rhoads favored continued individual allotment of Indian lands to further the goal of making the Indian a "self-supporting and self-respecting citizen." Collier favored corpo-

---

31. *The Friend* (Philadelphia), v. 42, no. 52 (8/21/1869), p. 413.
   Hagan, p. 111.
32. Larry E. Burgess, The Lake Mohonk Conferences on the Indian, 1883-1916. Dissertation. Faculty of History, Claremont Graduate School, 1972. pp. iii, 378.

rate control of Indian lands through tribal councils and opposed any further allotment or taxation. By 1931 Collier had withdrawn his support of Rhoads and renewed his criticism of the department. In spite of this conflict, Rhoads implemented reforms in education, health, and welfare suggested by the Meriam Report. Between 1928 and 1933 appropriations nearly doubled and the quality of service improved.[33]

These are only a few examples of the contributions Friends have made to organizations whose purposes are similar to those of the Indian Committee.

The Indian Committee of Philadelphia Yearly Meeting continues to nurture the oldest form of service, the personal visit. John Woolman, while examining the motives for his own visit to the Indians at Wyalusing, expressed for many the fascination of being personally involved, "that I might feel and understand their life and the spirit they live in."[34] From Friends' first visit to American Indians in 1658 to the present, through all the vicissitudes time has brought for both Quakers and Native Americans, the personal visit has remained a basic form of interaction and has provided a foundation for corporate actions.

---

33. Lawrence C. Kelly, "Charles James Rhoads (1929-33)" In: *The Commissioners of Indian Affairs, 1824-1977,* edited by Robert M. Kvasnick and Herman J. Viola (Lincoln: University of Nebraska Press, 1979), pp. 263-271.
34. John Woolman, *Journal and Major Essays,* edited by Phillips P. Moulton (New York: Oxford University Press, 1971), p. 127.

Friends in the Delaware Valley

Bibliographic Suggestions

Battey, Thomas C. *The Life and Adventures of a Quaker Among the Indians* (Norman: University of Oklahoma Press, 1968).

The Case of the Seneca Indians in the State of New York. Illustrated by facts. Printed for the information of the Society of Friends, by direction of the joint committees on Indian affairs, of the four yearly meetings of Friends of Genesee, New York, Philadelphia, and Baltimore (Philadelphia: Merrihew and Thompson, 1840).

Hubbard, Jeremiah. *Forty Years Among the Indians: a descriptive history of the long and busy life of Jeremiah Hubbard* (Knightstown, Ind.: The Bookmark, 1975).

Kelsey, Rayner Wickersham. *Friends and the Indians, 1655-1917* (Philadelphia: The Associated Executive Committee of Friends on Indian Affairs, 1917).

Parrish, Samuel. *Some chapters in the History of the Friendly Association for Regaining and Preserving Peace with the Indians by Pacific Measures* (Philadelphia: Friends Historical Association of Philadelphia, 1877).

Penn, William. *William Penn's own account of the Lenni Lenape or Delaware Indians.* Rev. ed. Edited, and with an introduction by Albert Cook Myers (Somerset, N.J.: Middle Atlantic Press, 1970).

Simms, Ruthanna. *As long as the Sun Gives Light: an Account of Friends' Work with American Indians in Oklahoma from 1917 to 1967* (Associated Executive Committee of Friends on Indian Affairs, 1970).

Some account of the conduct of the Religious Society of Friends towards the Indian Tribes in the settlement of the colonies of East and West Jersey and Pennsylvania: with a brief narrative of their labours for the civilization and Christian instruction of the Indians, from the time of their settlement in America, to the year 1843. Published by the Aborigines' Committee of the Meeting for Sufferings (London: Edward Marsh, 1844).

Tatum, Lawrie. *Our red brothers and the Peace Policy of President Ulysses S. Grant* (Philadelphia: John C. Winston, 1899).

White, Barclay. *The Friends and the Indians. Report of Barclay White, late Superintendent of Indian Affairs in the Northern Superintendency, Nebraska. Exhibiting the progress in civilization of the various tribes of Indians whilst under the care of Friends as agents* (Oxford, Pa.: Published for the Convention, 1886).

VI-B   The Yearly Meeting and Japan
by ELIZABETH GRAY VINING

Introduction

When in 1923 the Mission Board was established by Philadelphia Yearly Meeting (Arch Street), Friends' work in Japan already had a history of nearly thirty-eight years. The Women's Foreign Missionary Association of Friends of Philadelphia had been started in December 1882, under the leadership of Mary H. Morris; it quickly expanded, with branches in local meetings. During their first few years they sent $350 a year to a North Carolina Friend who was working in Mexico under Indiana Yearly Meeting, as well as smaller sums to a mission of English Friends in India and to the work for women and girls in Ramallah, Syria, which had been inspired by Eli and Sybil Jones of Dirigo, Maine.

In 1883 a Japanese student in Philadelphia wrote the WFMA that there was an open door for Friends' principles "among the higher classes" in Japan. A burning desire on the part of the Japanese to learn English was opening the way for educational work.

Two years later the WFMA raised $1500 to send their first missionaries, Joseph and Sarah Cosand, of Kansas Yearly Meeting, to Tokyo. In October 1887, the Friends Girls School was started in temporary quarters with three girls.

As the work in Japan grew and costs became heavier, they dropped the word *Women's* from their title, and in 1898 they became the Foreign Missionary Association of Friends of Philadelphia, including

Friends in the Delaware Valley

such liberal men as Jonathan P. Rhoads, Asa Wing, Joseph Elkinton, and others. Joseph Elkinton had a special interest in Japan, for his sister Mary had married Dr. Inazo Nitobe, a leading Japanese educator, and was living in Japan. Inazo Nitobe, as long as he lived, was an adviser to the Mission, from his first advice in 1890, when he told the WFMA that in selecting men to go to Japan they must have "choice men, the very best." The earliest reports used the word *heathen* rather freely, but after they met Dr. Nitobe, the word disappeared completely, as well as the attitude which it reflected.

In 1919, when the work had grown still further and the Student Christian Movement was drawing Young Friends into an interest in foreign missions, the FMA turned to the Yearly Meeting for recognition and support.

Establishing the Mission Board

The Yearly Meeting did not move precipitately. It appointed a Committee on Christian Labor in Foreign Lands, which consisted of twenty Friends, including some members of the FMA, to spend three years considering the question. In 1922 these Friends made a favorable report, recommending the establishment of a Board or Committee as an intrinsic part of the Yearly Meeting, which should consolidate the foreign work not covered by the AFSC. The Yearly Meeting, however, still felt that it did not yet know enough to act and therefore appointed another committee to confer with the FMA and bring in a plan.

Perhaps the delay was due to a certain ambivalence among Philadelphia Yearly Meeting Friends as to the fundamental question: Should Friends send out missionaries at all? If the Light Within is universal and "lighteth every man that cometh into the world," then should Friends impose on people who already have an ancient and satisfactory religion of their own, an alien religion, however superior it might appear to us? The Committee that brought in the final report in 1923, however, had no doubts about the importance of missions. "We wish to state as fundamental our conviction that Christianity is a universal and an essentially missionary religion and that in so far as Quakerism realizes its early designation, 'primitive

Christianity revived,' it too will be driven by its world vision to the uttermost parts of the earth."

The Yearly Meeting accepted the recommendation of the Committee that a Mission Board be established as one of the standing committees of the Yearly Meeting. It was to consist of an Executive Board of fifteen persons (later eighteen), a General Board made up of members of the Executive Board and two members appointed by each Monthly Meeting, and local boards in monthly meetings. (This elaborate structure was to be gradually simplified.) "The work in Japan," pronounced the Committee, "must be the first interest of the new Board."

It was a flourishing mission when the Yearly Meeting assumed responsibility, and the transition was made without interruption. The new Board was composed of many who had been members of the old FMA. In 1924 it made its first report.

The Work in Japan, 1924–1930

Lloyd Balderston, who had in 1921 returned from teaching at Hokkaido Imperial University, was chairman of the Executive Board, Dr. Edward G. Rhoads vice-chairman and his wife, Margaret W. Rhoads, secretary. These Friends had children already working in the Mission in Japan: Esther Balderston Jones and Esther B. Rhoads. Other Philadelphia Yearly Meeting Friends in the Mission were Edith Sharpless and Herbert and Madeline Nicholson.

During the twelve months since the establishment of the Mission Board two important events had focussed the attention of Americans and especially of Friends upon Japan: the Great Earthquake of September 1923, and the threat of the passage of the Johnson Immigration Bill, later known as the Oriental Exclusion Act. Money ($32,000) was forwarded through the AFSC to help in the relief and rehabilitation efforts of our missionaries and the newly formed Japan Service Committee, notably the building of a model village in Shiba Park, Tokyo, for twenty-seven families who had been burned out in the fire that followed the earthquake. Damage done to the buildings of the Mission itself amounted to $10,000.

In the United States, while the Johnson Act was still pending, two weighty members of the Yearly Meeting, William B. Harvey, who was on the Mission Board, and Francis R. Taylor, had gone with Gilbert Bowles, then in the United States on furlough, to labor with United States senators in a vain attempt to prevent the passage of this disastrous bill.

"One political event of the past year has been of paramount importance to all missionary work in Japan, namely the passage of the Johnson Immigration Bill with its exclusion clause," said the Report of 1924. "With justice the Japanese have felt this tacit abrogation of the Gentlemen's Agreement, hitherto so faithfully observed by them, to be a slur on their honor and have interpreted our refusal to admit them on a quota basis as an insult to their race out of all proportion to its slight effects in actual reduction of immigration."

Though the number of workers in the Mission varied from year to year, as some went home on furlough, others retired and new ones came on short assignments, there was a nucleus of Friends who remained in Japan for longer periods and were essentially the backbone of the Mission.

The heads of the Mission were Gilbert and Minnie Bowles. In 1893 Minnie Pickett of Glen Elder, Kansas, after a year's special work at Friends Select School, had been sent by the WFMA to work in Tokyo. Five years later, back in the United States on furlough, she married Gilbert Bowles of North Branch, Kansas, and in 1901, with their first child, Herbert, they went to Tokyo to head up the Mission, Joseph Cosand having left Friends and joined another denomination. In the twenty-three years since then Gilbert Bowles had won great respect among Japanese leaders, and among other missionaries in Japan, for his outstanding work for peace, temperance and education.

The activities of the Mission were centered in Tokyo and in Ibaraki Prefecture to the northeast, which was the area allotted to Quaker work by agreement among the Protestant churches.

In Tokyo there were two centers within a quarter of a mile of each other. One, in Daimachi, contained the residence in which Thomas and Esther Jones lived and which some years later became the Bowleses' home, a men's dormitory for students of nearby Japanese universities, and the home of Seiju Hirakawa, principal of the Friends Girls School. The other, Koun-cho, held the Bowleses' house, the Friends Girls School and its dormitory, the meeting house, a kindergarten, a small house for the two foreign teachers, and the Friends Institute for Community Work, where classes for women in Bible, English and Cooking were held.

In 1911 the Friends Girls School had been recognized by the Japanese Government and was known as one of the best girls' schools in Tokyo. By 1924 it had an enrollment of approximately 250 girls and 231 living graduates. The School was managed by a Board of Trustees, made up of about equal numbers of Japanese and Americans, but more than half of the income came from Japanese sources, chiefly parents and alumnae.

Esther B. Rhoads, though only twenty-seven in 1924, was already a leader in the School and the Institute. She had first gone to Japan in 1917 for a year, had returned to the United States to take an A.B. at Earlham in 1921, and then had returned to the Mission in Tokyo with a lifelong commitment. She was a lovely young woman with wavy brown hair, strong and enthusiastic and dedicated from childhood to working in Japan. She taught English, Cooking, and Bible both at Friends Girls School and at the Institute, and was head of the dormitory for forty-nine girls which the School then maintained.

Edith Newlin, of Iowa Yearly Meeting, who had taught in the School since she came to Japan in 1918, was in 1924 in the United States on furlough.

Thomas E. Jones of Indiana Yearly Meeting, his wife Esther and their two small sons were living in the new house in Mita Daimachi. Thomas Jones taught Economics at Keio University, while Esther Jones was homemaker and hostess. George Burnham Braithwaite, grandson of the greatly respected English Friend, J. Bevan Braithwaite, was at this time secretary and treasurer of the Mission. He had recently married Edith Lamb of Dublin Yearly Meeting in the new meeting house at Koun-cho. Their work was supported by the British and Foreign Bible Society, with which his parents had served.

In Ibaraki Prefecture American Friends were working mainly in two towns, Mito and Shimotsuma, though Japanese Friends were active in two or three others.

Edith Sharpless, daughter of Isaac Sharpless, President of Haverford College, had gone to Japan in 1910 and had settled in Mito, an old castle town and a citadel of nationalism, where there had been a good deal of opposition to Christian missionary effort. A dedicated and much loved worker, she ran a kindergarten for thirty-two children and a girls' dormitory, gave encouragement to the WCTU and conducted Bible classes. She had early achieved a remarkable command of the Japanese language.

Also in Mito were Herbert and Madeline Nicholson and their three small children. They had built a house, in which they welcomed all who came, conducted Bible classes and encouraged various local enterprises. Herbert taught English in local government schools.

Gurney Binford, who had been in Japan since 1893, and his wife Elizabeth were supported by Canada Yearly Meeting. They were settled in Shimotsuma, where they ran a kindergarten and worked with the country people round about. The Friends Meeting which they started was the only religious center in the town.

Though not all of the Quaker missionaries in Japan were members of Philadelphia Yearly Meeting or supported by it, they were all counted as part of the Mission.

The remaining years of the 1920's went off smoothly and the work flourished. In 1925 the budget for Japan was $25,289. Of that, travel and maintenance for missionaries on furlough amounted to $5,500, and the home office was run at a cost of $1,000. The school at Ras-el-Metn, Syria, received $500.

The report in 1926 set forth the philosophy of the Mission Board in notably liberal terms. "The second great purpose which the Mission Board is trying to accomplish"—the first one being "the extending the knowledge of Christ to the people of Japan"—"is the discovery of a policy of mutual help which shall gradually come to rule our relations as Christians with Christians of other countries . . . Every officer of Japan Yearly Meeting is a Japanese except an assistant treasurer. Japan Yearly Meeting is in full control of its own extension work. The head of Friends Girls School [Seiju Hirakawa] is a Japanese, which is unusual among mission schools in Japan, and the alumnae are taking a great interest in the school . . . We look forward to a time when they [the Japanese] will be able to help us by sending their spiritual ambassadors to America." At this time fifteen Japanese Friends were giving all their time to mission work as evangelists, teachers or assistants to the missionaries.

This was six years before the Laymen's Committee Mission

The Yearly Meeting and Japan

Report, *Re-Thinking Missions,* urged Christian missions in India, China, Burma, Korea, and Japan to give more scope to indigenous churches.

The Friends Girls School was pronounced by the Mission Board "worthy to be numbered among Yearly Meeting schools." (What higher praise could there be?) It not only had a Japanese principal but when Seiju Hirakawa resigned in 1927 to become Secretary of Japan Yearly Meeting, the School had a woman principal, Toki Tomiyama. She herself had been educated in the School and then, after a year at Westtown to perfect her English, had taken a degree at Teachers College, Columbia.

In November 1929, Princess Chichibu, the Emperor's sister-in-law, visited Friends Girls School, a signal honor for which impressive preparations were made. As Setsuko Matsudaira she had, when her father was ambassador to the United States, attended Sidwell Friends School in Washington.

Missionaries came and went. Thomas and Esther Jones resigned in 1926 to go to Nashville, Tennessee, where Thomas Jones would be President of Fisk University. They were well prepared, commented the Report, for work in "inter-racial adjustment." Esther Rhoads returned to the United States for the academic year 1926–1927 to take a master's degree in Religion at Columbia University.

Japan Yearly Meeting had developed a strong peace testimony. It was interested also in the cause of temperance and in the prevention and treatment of tuberculosis. A new meeting house in Tsuchiura was built largely through the efforts of Japanese Friends.

International understanding was an important emphasis of the Mission. "Gilbert Bowles," it was said, "is known where men of God will meet in many parts of the Far East."

The 1930s

The following decade was a difficult one and brought changes. Because of the Great Depression in the United States, the death of several large contributors and Japan's aggression in Manchuria and China, contributions fell off seriously. By 1934 there was a reduction of 61 percent in annual contributions to the Mission Board.

Friends in the Delaware Valley

"The deplorable happenings of the last months in Manchuria and Shanghai and Nanking," said the Report of 1932, "have subjected peace lovers in Japan to a severe strain. Let us not lose faith in the good will of the Japanese people, however much we must regret the resort of the military to violent methods in China." Two years later there was an even more ominous note: "The tide of life is rolling in against the small group of Christians in Japan."

J. Passmore Elkinton became Chairman of the Mission Board when Lloyd Balderston died in 1933. In that same year Margaret Rhoads also died. Both had given generously of themselves to the work of the Mission in Japan.

Throughout the decade Friends travelled both ways. In 1938 it was reported that since 1890, when Mary H. Morris and Mary H. Haines had visited Japan "in gospel love," ninety-eight members of Philadelphia Yearly Meeting had gone to Japan, some of them more than once. A number of Japanese Friends came to the United States. Most of them spent some time at Pendle Hill, which had opened its doors in 1930.

Kikue Kurama, at Pendle Hill in 1934 on the invitation of Young Friends and Pendle Hill itself, commented sadly, "The atmosphere of warlike preparations between our two countries is very discouraging." She made many talks to Friends groups and it was said of her, "She shows plainly the point of view of a second generation Christian. Her insight and deep concern for living out the implications of the Quaker way of life is stimulating to us."

Workers in the Mission were now going for one or two years only, with the exception of Esther Rhoads, the Bowleses, Edith Sharpless, The Nicholsons, and the Binfords. Non-Japanese teachers at the School on one or two year appointments included Luanna Bowles, a niece of Gilbert Bowles, Sarah A.G. Smith of Germantown Monthly Meeting, and Helen Thomas of Lansdowne Meeting.

1937 was the year of the Friends World Conference in Swarthmore and also the fiftieth anniversary of Friends work in Japan. Three members of Japan Yearly Meeting, Seiju Hirakawa and his wife, of Tokyo, and Ryumei Yamano of Mito came to the United States for the Conference, Ryumei Yamano to stay on for a year at Pendle Hill. Four members of Philadelphia Yearly Meeting, Robert H. Maris, and Thomas C. and Ethel B. Potts and their daughter Sarah, went to

Japan for the anniversary celebration. They found the meetings in Tokyo from October 16 to 18 very impressive and the Bowleses' house "a veritable Friends Center."

The following year the Mission Board "in view of the serious international situation" asked the Yearly Meeting for a grant of $400. As the Bowleses would soon be retiring after thirty-eight years of service, the Board wished to add a young couple to the Friends Center. They wished also to keep two American teachers on the staff of Friends Girls School and maintain the flow between the United States and Japan by an exchange of visitors. The Yearly Meeting, while expressing "especially sympathetic interest," encouraged Friends "to meet this request generously through their individual contributions to the Board."

The Board received $1742.31 from endowment fund income, the rest from contributions. Their total expenditures for the year ending in March 1938 were $14,998.93. They had an accumulated deficit of $5,065.95, which had to be made up from non-designated invested funds. A great deal was being done in Japan at remarkably small expense to American supporters. Four hundred dollars seems little to be asked for—and a good deal to be denied.

The Report in 1939 was even gloomier. "The Quaker group in Japan is going through a period of testing and sifting by war, mobilization, floods, earthquakes, requisitions, in faith, strength, resources." George Fox's *Journal,* Rufus Jones's *The Faith and Practice of the Quakers* and Howard Brinton's *A Religious Solution to the Social Problem* were being translated into Japanese. Aid was being given to sufferers from floods in Ibaraki Prefecture.

Contributions were $1,000 less than they had been the previous year. Herbert and Madeline Nicholson, when funds ran short, moved from Mito to Kobe and took positions in the Canadian Academy there.

Both Japanese and American Friends were greatly concerned about the plight of Friends in war-torn China. "Efforts to reach across in Christian fellowship to Friends in China have not been entirely unavailing," was the cautious report, "and are pursued at considerable risk." C. Walter Borton, a member of the Mission Board, and Gilbert Bowles visited China in 1938, Esther Rhoads and Minnie Bowles in 1939.

War clouds grew darker. There were shortages of all kinds in Japan and the Japanese were feeling the weight of world criticism. But the Friends Girls School had 530 students. Japanese Friends enlarged the Friends Center in Tokyo, building a new meeting house. A new Friends Meeting had been started in Osaka by the blind Friend, Takeo Iwahashi, whose book, "Light from Darkness" had been published in Philadelphia by the Yearly Meeting Book Committee. The Japan Yearly Meeting was planning to establish an annual lecture, similar to the Swarthmore Lecture, to be called the Nitobe Lecture.

In Philadelphia a newly established Japanese American Fellowship was active. About fourteen Japanese attended a weekend house party at Walter Borton's camp near New Lisbon on the Rancocas. An annual picnic was instituted that continued until 1968.

Doors were closing. Herbert and Madeline Nicholson and their children returned to the United States. Esther Rhoads, who had been at home on furlough, was not permitted by the United States government to go back to Japan in 1940. In August 1941, the Bowleses retired after more than forty years of service in Japan. They had been, the Report felicitously said, "cultural ambassadors rather than missionaries." Philadelphia Yearly Meeting commented, "These dear Friends are being sent back again as Japan's ambassadors to us." They settled in Honolulu, where they were to be of great service to the Japanese-Americans there throughout the War.

In 1940 the Japanese government had decreed the formation of a national church organization, to be called the Church of Christ in Japan, in which, for better supervision, the Christian denominations should be united. Though the Roman Catholics, the Episcopalians and the Holiness Church managed to stay out of it, Japanese Friends, after some hesitation, joined the new group. Possibly because of their very small numbers they felt more vulnerable than the larger denominations. The creedal definition for the new church organization was broad, but observance of the sacraments was required; pastors were therefore a necessary element. Friends found this difficult, but they submitted. They continued to gather in their meeting houses, holding a programmed service with a pastor, until the bombing raids of 1945 destroyed the Meeting House in Tokyo and about half of those in Ibaraki.

The Yearly Meeting and Japan

A few young Friends, however, stayed out of the new Church of Christ, meeting—at some risk—first in the Friends Dormitory in Daimachi and later in private houses. Harry Silcock, an English Friend who had spent many years in China, visited Japan for two weeks in 1941. He wrote: "We have yet to learn what sort of fellowship will be worked out by this small 'remnant.' "

The last pre-war letter from Japan Yearly Meeting, dated August 25, 1941, was read to Philadelphia Yearly Meeting in March 1942, when the war was already three months old. The letter mourned the departure of the Bowleses. "These dear Friends are sent back again as Japan's ambassadors to the United States," thus echoing what the Friends in Philadelphia had already said of them.

The War Years

Edith Sharpless was now the only American Friend still in Japan. She had moved from Mito to Tokyo and was living in the Bowleses' house, where she continued to give Bible lessons. She taught in the Friends Girls School for several months, but resigned in the summer of 1942. Japanese Friends loyally supported her until, in December 1943, she was repatriated to the United States.

The Mission Board, now cut off from Japan, set itself to counteract the deadly prejudice against the Japanese-Americans, many of them born in this country, that was boiling up on the West Coast. Former missionaries living in California began to gather information about the needs of the Japanese there and to make friendly calls. Alice Lewis Pearson, who had taught at Friends Girls School from 1905 to 1924, Gurney and Elizabeth Binford, Herbert and Madeline Nicholson, joined by Esther Rhoads, who went from Philadelphia for the purpose, worked long hours among distressed Japanese. When the relocation order went out, which tore them from their homes and businesses and confined them in concentration camps, the Mission Board and the AFSC worked together to help victims of this "forced migration." As young Japanese were gradually released from the camps when jobs were found for them on the East Coast, the Mission Board assisted those in the Philadelphia area, who in 1943 numbered 130, in 1944, 300 and in 1945, 500.

Friends in the Delaware Valley

In 1944 the Mission Board began to question itself as to whether it should ask to be laid down. The leaders of the Christian movement in Japan were capable men and women. Denominations—it seemed then—were gone. What place could there be now for a Friends Mission? But still they understood that the Friends Girls School and the Friends Center in Tokyo were going on under Japanese management. Might they not be foundations for future Quaker work when—some day—the war would be over? There was, they believed, still that small Quaker "remnant" that met in silent worship. They knew, moreover, that friendships with individual Japanese would still be firm and that after the war there would be a need for reconciliation among all groups. The Mission Board accordingly asked for an appropriation of $500 for the next year to continue their "testimony for racial equality."

When, on March 27, 1945, there was a joint session of the two Philadelphia Yearly Meetings at Arch Street, it was clear that the war would be over before many months. The Meeting wrote a letter to Friends in Japan, "to be sent when possible." "In deep penitence," it said, "for our share in your suffering and out of great love for you we pray that together we may be used by Him in rebuilding a world wherein dwells righteousness and peace and whose members are brothers and sisters."

The Mission Board at that meeting defined two lines of work: extending aid and friendship to the people of Japanese ancestry in this country, and being ready to re-open relations with Japan as soon as they became possible.

It was a hopeful meeting, that day in March 1945. Race Street and Arch Street Friends had met together on this important project and the end of World War II was in sight.

The Return

A year later Esther Rhoads was preparing to return to Japan as representative both of the Mission Board and of the AFSC. Mail from Japan was still strictly limited by the Occupation but occasional messages and letters came via friendly G.I.'s. A letter had come from young Friends in Japan, the "remnant" of whom Harry Silcock had written in 1942: "We feel that the future of Friends in Japan rests upon our shoulders."

When Esther Rhoads reached Tokyo in June 1946, she found the School, the Meeting House and the other buildings in the Koun-cho compound utterly destroyed. At Daimachi the Bowleses' house was still standing, though in bad repair, and there the young Friends were meeting. By this time they had attracted to themselves thirty or forty others.

Even a full year after the end of the War, the cities of Japan were devastated, though the sound of hammers was being heard everywhere. People were homeless, hungry and ill-clad. Licensed Agencies for Relief in Asia (LARA) of which the AFSC was a part, brought to them clothing, dried milk for the babies, and friendship. The story of Esther's work for LARA is told elsewhere. The extraordinary thing was that she was also, on behalf of the Mission Board, starting out to help rebuild the School and serving as a beacon light to the revived Tokyo Meeting. She wore her two hats with strength and wisdom and grace.

In the following March the Mission Board reported Esther Rhoads's work to the Yearly Meeting. It added also that Gilbert Bowles was in Japan on a five months' visit, that Luanna J. Bowles was working in the Civil Information and Education Department of the Occupation, and that "Elizabeth Gray Vining, a member of this Yearly Meeting and under appointment of the Imperial Household as tutor to the Crown Prince, is a Quaker emissary in her own right."

Friends Girls School

In rebuilding the School Esther Rhoads had, of course, the help of devoted alumnae and teachers. The first classes after the War were held in a battered concrete building in the center of a large burned-out area. Sheets of green plastic took the place of glass—which was unobtainable—for windows, and the girls perched uncomfortably on little wooden stools. Funds were raised in Japan and Philadelphia and in September 1948, the first new building was ready for use. It was a small frame building on the grounds at Koun-cho, containing a principal's office, a teachers' room and two large classrooms. On

Sundays it was used for the Meeting for Worship. Now the first American teacher since the War, Lillie Roudabush, a member of Florida Avenue Meeting in Washington, D.C., went to teach English Conversation in the School. She lived at the Friends Center and also taught Bible classes there.

In 1948 Esther Rhoads became principal of the School, succeeding Toki Tomiyama, who retired after more than twenty years of teaching there. Three years later, after Elizabeth Vining had returned to the United States, Esther Rhoads, in addition to all the other things she was doing, assumed some of her work with the Crown Prince and other members of his family. She was greatly appreciated by the Imperial Family.

The second unit of the School was completed in 1949. There were now 500 girls crowded into the two buildings. A third building was finished two years later and in 1956 a science building was added. The physical plant was now considered to be 60 percent rebuilt. The first wooden building had been replaced by a concrete one like the others, and further buildings, gymnasium, auditorium, a small apartment house for teachers, came along in due course.

Two and occasionally three American teachers taught English Conversation. They went to Japan on two year appointments and they were selected and supported by the Mission Board. Janice Clevenger of Bloomington (Indiana) Meeting was the only one to stay as long as ten years.

Japan Yearly Meeting

In November 1947, the first large gathering of Japanese Friends since the War took place at the Friends Girls School in Tokyo. Friends from the three revived monthly meetings, Tokyo, Mito, and Osaka, were present. In two days they re-established the Japan Yearly Meeting of Friends, with Takeo Iwahashi of Osaka and Kyoshi Ukaji of Tokyo as clerks.

A little over four years later a new meeting house was built in the Daimachi compound. By 1958 the Yearly Meeting had 220 members. Of this small number, 5—3 men and 2 women—held important educational and governmental positions in Japan. In 1958, too, a new monthly meeting was added, the Toyama Heights Meeting.

The Yearly Meeting and Japan

This Meeting was to a great extent the result of the work of Thomas and Eliza Foulke, members of Race Street Yearly Meeting, who had spent the year 1949 in Japan under AFSC appointment. Much of their work had centered in the large housing development built by the city of Tokyo for citizens who had been burned out during the War. The houses were extremely small and crowded close together, but many of the occupants were college professors, business men, and other middle class people. A large building in the center of this settlement was acquired and in it the Foulkes helped to establish a neighborhood house with a nursery school, women's classes, and a library, which Violet Gordon Gray, then living with her sister, Elizabeth Vining, organized and catalogued. Eager to know what motivated the Foulkes, some of the residents became interested in Friends, and a Meeting for Worship was held regularly, as well as classes in Quakerism and the Bible. This Meeting became a regular monthly meeting in 1958.

Neil and Venette Hartman of Moorestown Monthly Meeting followed the Foulkes at Toyama Heights. Eliza Foulke was to return twice to Tokyo in later years, to serve in the Friends Center in Daimachi.

More Quaker Emissaries

It is difficult, if not impossible, to say at this juncture which Friends were workers of the AFSC and which of the Mission Board. If the AFSC paid their passage, the Mission Board housed them, and Esther Rhoads guided them. Missionaries under the Occupation were restricted in many ways; food, housing, and transportation were in short supply and obtaining them was a complicated process. Japan was still struggling to rebuild her devastated cities. Later, after the signing of the Peace Treaty, the work of the two Quaker groups would become more distinct, but there was always cooperation between them.

In 1950 Henry and Edith Perry of Cambridge (Massachusetts) Monthly Meeting came for seven months to be gracious and helpful members of the family at Friends Center.

The next couple (AFSC/Japan Committee) to come from Philadelphia Yearly Meeting were Howard and May Taylor, in 1950. They stayed a year this first time, living at the Tokyo Friends Center, working with Toyama and Setagaya Neighborhood Centers, seminars and work camps. They were sent again in 1958 by the Japan Committee to live in Shimotsuma and serve particularly in Ibaraki Ken, holding English classes, Bible classes, farmers' retreats, peace walks, cultural exhibits. They were endlessly hospitable and made many warm friendships.

In December 1950 Edith Sharpless had the joy of returning to Japan after seven years' absence. Back again in Mito, where a small addition to the meeting house had been built for her, she lived and taught, busy and appreciated, until she returned to Haverford in May 1956. In August of that year she died. She had given altogether forty-six years of service to Japan.

Howard and Anna Brinton were in Tokyo for two full and productive years, 1952–1954. Howard met with Japanese scholars and philosophers and taught classes at the Center; Anna helped run the AFSC/JFSC nursery school at Setagaya, was supervisor of the Friends Center during Esther Rhoads's absence on furlough, and made many friends. Both travelled widely about Japan.

Herbert Nicholson had returned to Japan in 1948, on behalf of Church World Service, bringing goats to Japanese farmers. In 1960 his son and daughter-in-law, Samuel and Anna Margaret Nicholson of Ann Arbor Meeting, came with their year-old son to work in Ibaraki Prefecture, where Samuel had been born. They settled in Shimotsuma first and later moved to Mito, where they constantly entertained visitors from the United States, Germany, England, Australia, New Zealand, and Korea. Samuel helped the AFSC with work camps and started ceramic classes at the Mito Old People's Home. They were there until 1966.

The first Japanese Friends to come to the United States after the War were Orie Shimazaki and Kiyoshi Ukaji, who were at Pendle Hill in 1948. Fifteen years later it was reported that there had been since the War twenty visiting Friends from Japan and Korea. Many American Friends also went to Japan on other errands than those of the Mission: as tourists, after travel became possible, as teachers in other institutions than the Friends Girls School, as members of learned societies and delegates to conventions. There had indeed

The Yearly Meeting and Japan

been, as Esther Rhoads wrote in 1947, "the miracle of endurance and the miracle of forgiveness."

Mission Board to Japan Committee to International Outreach Committee

In 1951, four years before the re-unification of the two Philadelphia Yearly Meetings, the Yearly Meeting (Arch Street) approved the reestablishment of the Mission Board as a joint committee with Race Street Yearly Meeting, to be called the Japan Committee of the Religious Society of Friends in Philadelphia, with representatives from both Yearly Meetings. It was Eliza Foulke's eloquent appeal to the General Meeting, telling how difficult it was to explain to the Japanese the fact of two Yearly Meetings in Philadelphia, that brought about—or at any rate hastened—their full merging in 1955. Eliza Foulke herself was one of the seven Race Street Friends appointed to the new Board.

The 1955 Financial Report of the Japan Committee in the Race Street Minutes is briefer and clearer than the Arch Street one: Income, $18,138.82. Expenditures: $19,018.71. Deficit, $879. It notes that besides the invested funds and contributions, Arch Street gave $1,000, Race Street $500.

After the Brintons returned from Japan, Anna Brinton succeeded William Cadbury as Chairman of the Japan Committee for the next three years, and was followed by Sarah Swan. Five years later, in 1960, Esther Rhoads, full of honors, including the Third Order of the Sacred Treasure, retired after more than forty years of distinguished service in Japan.

By 1960 the Peace Treaty had been signed, and Japan had become a member of the United Nations. Tokyo and the other large cities had risen out of their ashes and had been rebuilt, new industries were beginning to flourish and prosperity was returning. The Japan Yearly Meeting was preparing to take full responsibility for itself. During the twelve years from 1960 to 1972 there was a steady growth toward independence.

In 1964 the Japan Committee expenditure was $41,194.83, of

which half went for personnel, $15,186 as a contribution to a new auditorium for Friends Girls School, $3,857 to the Philadelphia Office, where Sylvan Wallen had been secretary since 1948, $1504 to Japan Yearly Meeting and Japanese visitors, and the rest to assistance for various small projects. Later in the decade there were also contributions to a small addition to the Friends Center and to the School, for some of which money had to be taken from bequests.

The American workers, except for the two teachers at the School, were steadily withdrawing. The Nicholsons returned to the United States in 1966. Fumie Miho, of Honolulu Monthly Meeting, who had been the American director of the Tokyo Friends Center since 1960, retired in 1967. In the Philadelphia Office in the next year, Sylvan Wallen retired. Sarah Swan became part-time executive secretary and Esther Rhoads chairman of the Committee. By 1967 Takeshi and Masa Watanabe, who had been studying at Woodbrooke and Pendle Hill, returned to Japan to become wardens of the Friends Center.

In 1971 the Japan Committee reported to the Yearly Meeting: "With the recognized ability of Japanese Friends to carry out their concerns and their diminishing need for the type of support provided in the past by this Yearly Meeting, the Japan Committee has begun to implement the changes in form and direction which were suggested a year ago."

Japan Yearly Meeting now comprised three monthly meetings in Tokyo, three in Ibaraki Prefecture, one in Osaka. Japanese Friends' special interests have been temperance, labor relations, education for women, education for the blind and mentally retarded, cooperatives, care for the aged, concern for peace. The Friends Center housed some permanent residents and many visitors, who came and went. The Friends Girls School was a first class six-year high school with 650 girls. Almost all of its graduates went on to college.

It took altogether five years to phase out the work of the Japan Committee. The Friends Girls School undertook full support of their American teachers, while the Friends Council on Education in Philadelphia would help to recruit and screen candidates. Japan Yearly Meeting had taken responsibility for the Tokyo Friends Center. The office of the Japan Committee was closed and its records sent to the Yearly Meeting Archives at Haverford College. Some of the endowment was transferred to the Friends Girls School, the rest to the Philadelphia Yearly Meeting for the use of a successor committee.

The Yearly Meeting and Japan

Ninety years after the Women's Foreign Missionary Association first began to think of Japan, Representative Meeting proposed to the Yearly Meeting that the Japan Committee be replaced by the establishment of a smaller committee with broader responsibilities, to be known as the International Outreach Committee. Initially it was to consist of three members from the former Japan Committee, three from the Friends World Committee and three younger Friends. The Yearly Meeting concurred with this proposal. The word *mission* had completely vanished, along with the overtones of meaning that it once conveyed.

The new Committee began its work with a meeting and potluck supper at the Moriuchis' house in Moorestown on May 4, 1974. Fifty-five people, including three members of Japan Yearly Meeting who happened to be in the United States at the time, and eight members of Philadelphia Yearly Meeting who had taught at Friends Girls School in Tokyo were there, as well as several Issei, Nisei and Sansei.

Funds are now shared with Friends in other countries besides Japan. In 1976, for instance, the income of $19,681 provided help to Quaker students from East Africa who were at American colleges, to a General Conference of Friends in India and to Malara Girls Secondary School in Kenya. There were also the continuing contribution of $600 to the Japan International Christian University Foundation, grants to Friends Girls School for help in rebuilding the gymnasium, and to Mito Monthly Meeting.

The International Outreach Committee is actively concerned with helping arrange and support intervisitation with Friends of other countries.

The WFMA, the FFMA, the Mission Board and the Japan Committee in their turn have served their purposes honorably and been devolved. The friendships and the understanding they have created live on.

The sources for this essay have been minutes, letters and reports of the successive committees from 1872 to 1976 in the Quaker Collection of Haverford College. I am grateful to Sarah C. Swan for reading my manuscript and correcting the errors that she found. Any errors that might remain are solely my responsibility.

VI-C   Philadelphia Yearly Meeting and the American Friends Service Committee

by MARY HOXIE JONES

The Special Peace Committee of thirty-one Friends appointed by Philadelphia Yearly Meeting (Race Street) at its sessions in 1916 reported that the committee's purpose had been to maintain and extend the testimony of the Society against unChristian methods of settling international problems by killing and destruction. The committee was in favor of sane and Christian methods of conference and goodwill. The minute stated:

> As directed by the Yearly Meeting, your committee has cooperated in the closest unity with a like committee previously appointed by the Arch Street Friends. . . . There have been joint meetings of representatives of the two committees . . ., frequent conferences of the chairmen of the sub-committees and constant interchange of service and of literature. . . . It is a great pleasure to report to this meeting (held May 14–18, 1917) the fine leadership, efficient service and cordial unity we have found among the Arch Street Friends in this mutual undertaking; and the experience emphasizes the sense of loss in the century of unnecessary separation of the two bodies with aims and ideals so closely in accord.[1]

Before the United States entered World War I on April 6, 1917, two large meetings had been held in the Arch Street and Race Street Meeting Houses, the first in November 1916 when Leighton Richards and Rufus M. Jones were the speakers (neither one a mem-

---

1. Philadelphia Yearly Meeting (Race Street) Minutes, 1917, pp. 138–139.

The Yearly Meeting and the A.F.S.C.

ber of Philadelphia Yearly Meetings); and the second when Isaac Sharpless, member of Philadelphia Yearly Meeting, Arch Street, spoke in February 1917 in the Race Street Meeting House.[2]

Not only were Philadelphia Friends concerned about the growing war clouds, Friends throughout the United States had been disturbed over threats to peaceful solutions. The Five Years Meeting, established in October 1902, in Indianapolis, had arranged for a Peace Conference to be held at Winona Lake, Indiana, in the summer of 1910. An outgrowth of this was the Friends National Peace Conference, meetings of which had been held in 1912, 1913, and 1915. These conferences were attended by members of both Philadelphia Yearly Meetings; Arch Street member Henry J. Cadbury was made chairman. Shortly before the United States entered the war, a meeting of this Peace Conference was held and issued "a message from the Society of Friends concerning the condition of affairs."

This message was printed as an advertisement in newspapers and magazines throughout the United States and contained these sentences on constructive service: "The alternative to war is not inactivity and cowardice. It is the irresistible and constructive power of goodwill."[3]

The Friends National Peace Conference called a meeting on April 30, 1917, and the small group included Friends drawn from both Race Street and Arch Street Yearly Meetings and the Five Years Meeting. These Friends were: Alfred G. Scattergood, temporary chairman; Charles J. Rhoads, Henry W. Comfort, Henry J. Cadbury, and Anne Garrett Walton from Arch Street; Jesse H. Holmes, Lucy Biddle Lewis, Arabella Carter, William H. Cocks, and J. Barnard Walton from Race Street; and L. Hollingsworth Wood, Homer L. Morris, and Vincent D. Nicholson from Five Years Meeting. They met together in the Young Friends Association Building, on Race Street property, and after this group of Friends finished their deliberations, the "Friends National Service Committee," its first and temporary name, was born.[4]

---

2. PYM (Race) Minutes, 1917, pp. 138–142.
3. Anna Brinton, ed., *Then and Now* (Philadelphia: Univ. of Pa. Press, 1960), pp. 23–24; Rufus M. Jones, *Service of Love in War Time* (New York: Macmillan, 1920), p. 7.
4. Mary Hoxie Jones, *Swords into Ploughshares* (New York: Macmillan, 1937), p. 319; R. M. Jones, *Service of Love in War Time,* pp. 8–9.

Various meetings of this group were held and on May 11 the name was changed to American Friends Service Committee. On June 1, Vincent D. Nicholson was made its executive secretary, a post he held for one year until he was drafted and later furloughed for work in France. He was given an office in the Friends Institute, 20 South Twelfth Street. On June 2, two Arch Street Friends, J. Henry Scattergood and Morris E. Leeds, sailed for Europe to investigate possibilities for relief work. On June 11 Rufus M. Jones, a member of the Yearly Meeting of Friends for New England, was made chairman of AFSC. On June 23 the first group of workers sailed for France—nine Friends from both Yearly Meetings: George V. Downing, Edith Coale, Douglas Waples, Eleanor Cary, Ernest L. Brown, Howard and Katherine Elkinton, William and Mary Elkinton Duguid.[5]

Haverford College provided facilities for a training unit as soon as the war started and this definitely was a step in facilitating the preparation for service abroad.

Although both Philadelphia Yearly Meetings had had peace committees for a number of years and the Service Committee might have been established as a merger of these two groups, it was far better to have it arise as a totally separate and new entity which encompassed the whole area of American Quakerism. However, the two Yearly Meeting Peace Committees were closest geographically to the AFSC and for a period of years they undertook a major amount of the fund raising, arousing interest and support from the constituent monthly and quarterly meetings of both Yearly Meetings. Funds were channeled through the Peace Committees to the AFSC.

The Yearly Meetings also appointed members to serve on the AFSC, long before the Board of Directors as now structured was in existence, or before the AFSC corporation was formed in 1928. AFSC meetings in the early days were made up of Philadelphia Yearly Meeting members, other yearly meeting members and any Friends who wanted to attend. It should be mentioned here that while the emphasis of this chapter is on the Philadelphia Yearly Meetings and the AFSC, it is essential to remember that nearly all American yearly meetings and groups of Friends played important parts in the AFSC. This aspect has not been forgotten.

---

5. *Swords into Ploughshares*, p. 319.

The Yearly Meeting and the A.F.S.C.

Naturally, the two Philadelphia Yearly Meeting felt extremely close to the AFSC, as indeed they were in fact. The AFSC meetings and their subcommittees met either in the Cherry Street room at Race Street, or in the Monthly Meeting room at 20 South Twelfth Street, and eventually both of the buildings at 20 South Twelfth Street, Friends Institute and Meeting House, were occupied by the AFSC. The small shed at Fifteenth and Cherry Streets served as "warehouse" for the supplies and clothing sent abroad until a large room at 1520 Race was secured, and in 1940 a large warehouse at Twenty-third and Arch Streets.

Though never technically a child of the two Yearly Meetings, AFSC was an adopted one, and without the properties of the two meetings, the backing, the personnel, and the financial assistance, the AFSC would have had a much harder time getting itself under way. The complications due to the war, the draft, and the red tape involved in getting men furloughed from army service to relief service, made the first year and a half, before the end of the war on November 11, 1918, a time filled with extraordinary difficulties.

The Committee on Peace and Emergency Service of Philadelphia Yearly Meeting, Race Street, reported to the Yearly Meeting in 1919: "Reconstruction has been carried on in close cooperation with the AFSC which represents all various bodies of Friends in the United States and links up as well with Friends in England . . . the greatest unifying influence that has been active among Friends since the unfortunate period of separations in the last century." The first year of the AFSC work was a tripartite organization with the American Red Cross and the British Friends War Victims Relief Committee. The "War Vics" later joined with the Friends Foreign Missionary Association to become Friends Service Council. During World War II the Friends Relief Service was established and in 1948 it was merged with Friends Service Council. This is now called Quaker Peace and Service, as of 1978.

The AFSC requested that the Race Street subcommittee on finance should receive contributions for reconstruction work from various monthly meetings and individuals. This committee worked closely with a similar committee of Arch Street Friends.[6]

---

6. PYM (Race) Minutes, 1919, p. 105.

Friends in the Delaware Valley

"When the United States entered the war and began to marshal its military and naval forces," the Arch Street Minutes recorded, "the Peace Committee did what it could, at first to prevent the passage of a draft law of any kind, and later when the act seemed inevitable, to modify its provisions . . . Since the Eleventh Month, when the responsibility for the drafted young men was assumed by the [Arch Street] Representative Meeting under the care of a special secretary [William B. Harvey] such action as we have taken has been in conjunction with him and the AFSC."

The AFSC originated "to meet the needs of American Friends in war time. A great vision of all Friends in this country standing together... if possible, for a common service, inspired its founders." From the beginning, the Peace Committee of Philadelphia Yearly Meeting (Arch Street) has had an active share in the creation and continuous development of the AFSC. Members represent the Peace Committee on the AFSC and share with representatives of other bodies of Friends in determining its policies.

The extraordinary work of AFSC cannot be described here. This is an account primarily of the relationship of AFSC and Philadelphia Yearly Meetings. There are many books, reports, and articles which give full accounts of the reconstruction, child-feeding and social industrial projects carried on during the years.

The Peace Committee organized campaigns for funds as well as sewing and knitting groups. Both at home and abroad it had been a period of growth.

> The very existence of a committee on which all American Friends can unite in a common service is in itself a step forward that hardly seemed possible one year ago. . . . When we consider the way in which those who were needed have been found to carry on the work, there is little doubt that the AFSC has been the expression of an irresistible impulse of the time. . . . We stand today amid the moral wreckage of a materialistic civilization. The task for Christians is to lift mankind to a new spiritual plane and rebuild social life on the one sure foundation of love.[7]

Another minute for this same year, 1918, stated that "the close affiliation with AFSC and the harmonious work carried on is gratefully appreciated. The appropriation of $1500 (to the Peace Committee to give to the AFSC) was granted."[8]

---

7. PYM (Arch) Minutes, 1918, pp. 66–75.
8. *Op. cit.,* p. 13.

The Yearly Meeting and the A.F.S.C.

By 1919 the Arch Street Peace Committee reported a large amount of sewing, knitting and canning had been achieved, an invaluable contribution from Friends.[9] The knitting and clothing groups of the different yearly meetings throughout the United States have provided the closest links for Friends and others with the AFSC, ties which have held all through the sixty-three years of the life of AFSC. By 1921 the two Peace Committees of the yearly meetings were working more and more closely together, and this cooperation was called "a work of peace tending to wipe out the evil passion which brought on the separation of nearly one hundred years ago."[10]

By 1923 funds coming from the monthly meetings decreased to some extent and five years later in 1928 there was sadness at the apparent lack of interest in supporting the AFSC. "AFSC's financial condition is truly serious. We must rally to support."[11] A year or two after this the AFSC started to have its own fund raising program. In 1933 the two Peace Committees of the two Yearly Meetings merged into one—the Friends Peace Committee of Philadelphia and Vicinity.[12] Its office was at 302 Arch Street, and the information center at 1515 Cherry Street. The next year the joint report was signed by a member of each Yearly Meeting.

Relief and emergency work abroad were decreasing and there was an urgent consideration of the AFSC's future. Reconstruction and child-feeding hopefully were ended forever. Did the future need an AFSC? It seemed unwise, however, to close it down. It was time to consider other forms of service needed in the United States, so in 1925 AFSC was organized in four sections: Peace, Foreign Service, Interracial, and Home Service. The chairmen of the last three were members of Race Street Yearly Meeting. The chairman of the Peace Section was soon to be a member of a new, united monthly meeting affiliated with both Yearly Meetings.

The AFSC Peace Section worked closely with the Peace Committee of the two Yearly Meetings. The latter was asked to participate in Institutes of International Relations; Peace Caravans were started in 1927. College students, in teams of two, drove model T Fords to outlying communities to talk about peace. The Swarthmore College

---
9. PYM (Arch) Minutes, 1919, p.39.
10. PYM (Race) Minutes, 1921, p. 97.
11. PYM (Race) Minutes, 1928, p. 55.
12. PYM (Race) Minutes, 1934, p.21.

Caravan, though arranged by AFSC, was financed by the Peace Committee and this arrangement continued for a number of years. For the next decade or more the Yearly Meeting Minutes note close cooperation with the AFSC in supporting pacifist groups, conferences, and the distribution of literature. The outbreak of World War II in 1939 made peace education all the more vital. Instead of being completely disheartened by failure to achieve permanent peace among nations, efforts by both AFSC and the Yearly Meetings were increased during the 1940's, 50's and 60's.

In the 1970's sit-ins, vigils and protests were organized either by AFSC or the Peace Committee, or both groups joined with other peace organizations in widespread campaigns:

> Stop the B-1: National Peace Conversion Campaign which George Lakey [secretary of the Peace Committee] is directing in 1974 in conjunction with the national campaign which is sponsored by the AFSC and Clergy and Laity Concerned [was] an effort to divert funds for B-1 to housing, health care and mass transit.[13]

In 1975 a member of the Peace Committee staff joined an AFSC visit to the Middle East to gather information as a basis for a Yearly Meeting educational program, which included speaking engagements.[14]

In 1977 the AFSC and the Yearly Meeting signed jointly a cable to President Park of South Korea urging release of the nine political prisoners and the restoration of full civil rights to others convicted. The draft of the cable had been prepared by the Peace Committee.

AFSC's Interracial Section, like the Peace Section, was closely affiliated with the Race Relations Committee of the two Yearly Meetings. In 1919 a Race Relations Committee was established in Arch Street when a member had a concern to see what could be done to abolish lynching. This concern was presented also to Race Street, where it met with sympathetic response.[15] Under their Philanthropic Labor Committee, Race Street had been working for the rights of colored people. In 1923, at a joint session of men and women Friends at Race Street, Elizabeth Powell Bond spoke feelingly, saying, "There is no place in the world where the vital question [of race relations] can be so fittingly taken up as in the Society of Friends."[16]

---

13. PYM Minutes, 1974, p. 169.
14. PYM Minutes, 1976, p. 72.
15. PYM (Race) Minutes, 1919, p. 32.
16. PYM (Race) Minutes, 1923, pp. 21–22.

The Yearly Meeting and the A.F.S.C.

By 1925, a Race Street Friend was chairman of the AFSC Interracial Section as well as the Yearly Meeting Committee on the Interests of the Colored Race, the name of which was changed in 1928 to Race Relations Committee.[17] By 1927 there was joint activity. "In conjunction with AFSC and Race Street Friends, a hearty response was given to an invitation from colored welfare workers in Philadelphia to join in a banquet to discuss local interests." There was also "a tour of Negro districts studying educational, social and business institutions."[18]

By 1928, the minutes indicate that the AFSC's work in race relations and that of the Yearly Meetings' committee were overlapping. At an AFSC meeting in February it was decided that its work in this field should be national rather than local and that it would not undertake activities within the Yearly Meetings' limits. In order to implement these decisions, the two Race Relations Committees united and employed Helen R. Bryan as secretary. She had been briefly secretary of the AFSC Interracial Section, and although not a Friend, she had a complete understanding of Quaker principles and ideals. She began her work for the Yearly Meetings' committee in October 1928 with an office at the AFSC headquarters, 20 South Twelfth Street.[19]

In spite of the fear of overlapping there was close cooperation between the Yearly Meetings and the AFSC. In 1932 the Race Relations Committee cooperated with AFSC in holding a conference in Germantown, "attended by about one hundred Friends to call together a small group of Friends who believe that the Society is not meeting adequately the demands of the American racial situation."[20] There was a strong feeling of appreciation by the AFSC for the Yearly Meetings' concern.

An outstanding achievement of 1933 was the Institute of Race Relations held at Swarthmore College, with Clarence E. Pickett, of the AFSC and Charles S. Johnson of Fisk University as directors—one of the first attempts made in this country to secure over an extended length of time the latest and most able thinking in this field. It was a remarkable experiment, for it was said that through this

---

17. PYM (Race) Minutes, 1926, p. 31.
18. PYM (Arch) Minutes, 1927, p. 72.
19. PYM (Arch and Race) Minutes, 1928-30.
20. PYM (Arch) Minutes, 1932, p. 113.

Institute "Friends have put into action one of the most significant and potent techniques ever involved in race relations."[21]

These Institutes continued and cooperation between AFSC and the Race Relations Committee remained close. They combined with the Young People's Interracial Fellowship to issue a statement deploring the wave of anti-Jewish attacks in 1939.[22] In order to avoid duplication of work done by AFSC and the Race Relations Committee, as well as other groups now active in Philadelphia, a joint committee was appointed.[23] There have been frequent references over the years to the cooperation between AFSC and the Race Relations Committee particularly in the field of integrated housing. The new development at Levittown presented possibilities and there was great concern that "it be organized on a democratic basis."[24]

In preparation for the All Friends Conference to be held at Oxford, England, in 1952, the appointed delegates from the two Philadelphia Yearly Meetings, the Race Relations Committee and the AFSC explored together how to make discussions as fruitful as possible during the time at Oxford.

The Minutes note, year after year, that the Race Relations Committee has worked closely with the AFSC and "worked to ease tensions when a Negro family moved into Levittown." In 1963 there was an urge to send a mission to Mississippi "in consultation with race relations of AFSC and other groups."[25]

The Social Order Committee of Arch Street Yearly Meeting was established in the sessions of 1917 "to consider the part which the Religious Society of Friends should take in the present day application of efforts to promote the Kingdom of God on earth, particularly as it relates to social, political and industrial conditions."[26] Earlier than 1917, Race Street Yearly Meeting had formed a committee, already referred to above, called Philanthropic Labor. In 1947 these two committees began working together, merging into one committee two years later in 1949.[27]

---

21. PYM (Arch) Minutes, 1934, pp. 111-112.
22. PYM (Arch) Minutes, 1940, p. 156.
23. PYM (Arch) Minutes, 1945, p. 22.
24. PYM (Arch) Minutes, 1952, p. 151.
25. PYM Minutes, 1964, p. 24.
26. PYM (Arch) Minutes, 1918, p. 78.
27. PYM (Arch) Minutes, 1949, p. 171.

The Yearly Meeting and the A.F.S.C.

The Home Service Section of the AFSC was formed in 1925 and eventually became the Social Order Section in 1935—this had been urged by the two Yearly Meetings. The AFSC had begun its work camp program in 1934, and in 1935 the Social Order Committee helped the AFSC develop a work camp at the Bedford Street Mission in Philadelphia.[28] In 1939 David S. Richie began his service as secretary of the Social Order Committee where he continued until his retirement in 1973, devoting much of his time to work on AFSC projects as well. Two years after he came to the committee, the two committees of Arch and Race merged.

In 1945 the Social Order Committee and the AFSC co-sponsored a ten week Student-in-Industry project which was held at the College Settlement in Philadelphia and developed into a year-round student-in-industry program.

Ten years after the founding of the American Friends Service Committee, the Young Friends Committee of Arch Street Yearly Meeting, the Young Friends Movement of Race Street Yearly Meeting, and the Young Friends of the Five Years Meeting requested representation on the AFSC. This resulted from the Young Friends Conference held in the summer of 1927 at Oskaloosa, Iowa. "The concern was felt that the AFSC and the Young Friends ought not to be separate and distinct groups but that the two would gain by coming closer together. The result has been the appointment of Young Friends to the Executive and Regional committees of the AFSC."[29]

The Young Friends movement then had members upon the executive and all regional committees[30] and this arrangement continued for a number of years. By 1930 the two Young Friends groups of Philadelphia had become the Young Friends Movement—the first of the yearly meeting committees to unite. The AFSC financed a young Friend, Amy E. Sharpless, in 1928-1929 as a travelling field secretary for the Peace Section. This was done in cooperation with the three Young Friends Committees of Arch, Race, and the Five Years Meeting.

The first written report of AFSC was printed in the 1929 minutes of both Yearly Meetings. When joint sessions of the two were held,

---

28. PYM (Arch) Minutes, 1936, p. 137.
29. PYM (Arch) Minutes, 1928, p. 75.
30. PYM (Race) Minutes, 1928, p. 82.

Friends in the Delaware Valley

the report was generally presented then; when Philadelphia General Meeting was started in 1946 (its last session was in 1954), the reports were sometimes made then. In 1943, during the period of the Second World War, "gratitude was expressed for the work of the AFSC which is carried on so effectively on behalf of our whole Society."[31] There was great interest and concern for the Civilian Public Service Camps and the Detached Service projects in which many Yearly Meeting members were involved. A large amount of fund raising was undertaken by the constituent monthly and quarterly meetings of the two Yearly Meetings.

The Fellowship Committee developed in 1932 from the Message Committee of AFSC. It grew out of the desire to increase intervisitation between meetings, both in the United States and abroad and to provide Friends groups with a wider knowledge of one another. In 1936 it merged with the Fellowship Council which had had as its function the nourishing of the life of the new independent meetings for worship "some sixty of which are carried on in various parts of the country, the stimulating of itinerant ministry and visitation."[32] By 1940 the Fellowship Council consisted of ninety-one members, chosen by the Council, approved by the AFSC, and including members of both Yearly Meetings. An invitation was made to the Yearly Meetings to appoint three official members to the Council.[33] By 1953 both Yearly Meetings approved a plan to amalgamate the Friends World Committee (American Section) and the Friends Fellowship Council. The new body was called the Friends World Committee for Consultation, American Section and Fellowship Council, with two members to be appointed from each Yearly Meeting.[34]

In the late 1920's a committee was formed to consider a more efficient use of Friends' properties in Philadelphia. There was a hope that one of these properties, Friends Select School, would move from its location at Seventeenth and the Parkway into the suburbs, thus making available land for a Friends Center. This did not happen and the hope for a geographical coming together of Friends activities did not occur at this time. By 1953, Clarence E. Pickett, executive

---

31. PYM (Arch) Minutes, 1943, p. 51.
32. PYM (Arch) Minutes, 1936, pp. 40-41.
33. PYM (Arch) Minutes, 1940, p. 54.
34. PYM (Arch) Minutes, 1953, p. 29.

The Yearly Meeting and the A.F.S.C.

secretary emeritus of AFSC, spoke in Yearly Meeting, saying that he hoped Friends' attitude toward property would be a manifestation of faith in the future and that there would be a Quaker attitude of responsibility to bear witness in a large city to the truth that we hold precious.[35]

Committees were formed; the city condemned property in order to widen Fifteenth Street between Race and Cherry Streets, where the AFSC was housed (after leaving the property at 20 South Twelfth Street in 1960) and plans were underway for a new building, situated some twenty feet farther in from the pavement on Fifteenth Street. In 1966 Representative Meeting proposed that the Yearly Meeting, Central Philadelphia Monthly Meeting, and the AFSC each appoint three persons to form a committee of nine to plan for a Friends Center. This was done and AFSC agreed to give $9,000 as a start.

The problems were complex and delays occurred. Many plans were drawn and submitted to the three groups involved. By 1973 it was reported to the Yearly Meeting that the basic relationship with the three groups was being established, that the Friends Center Corporation was officially organized on October 17, 1972, with nine members, three from each group. Later the Board was increased to twelve members—three to rotate as representatives of other Friends organizations housed in the new center complex. The AFSC offices were to occupy the second and third floors of the new building, the entrance to be on Cherry Street. The material aids department, moved down from Twenty-third and Arch Streets, would be housed in the renovated basement of the Race Street Meeting House.[36]

The AFSC moved into its new location, after being housed temporarily in the Patriots Building at Sixteenth and Sansom Streets, on June 30, 1975, a week before the Yearly Meeting and other groups moved in. The Friends Center was dedicated on November 13, 1975.

In spite of this geographical nearness, the AFSC and Philadelphia Yearly Meeting found themselves less close together in their activities, as both groups carried on full programs of their own. At the Friends General Conference in 1979, held in Richmond, Indiana, there was considerable expression that the AFSC was not sufficiently responsive to the Society of Friends, nor did it always follow Quaker

---
35. PYM (Arch) Minutes, 1954, p. 8.
36. PYM Minutes, 1973, pp. 210–211.

principles. A special study of these serious criticisms was undertaken during 1979-1980 by a special committee appointed for the purpose. It was hoped that the geographical nearness in the Friends Center would facilitate this study. A few sentences from the report made to the Board of AFSC in June 1980 are pertinent:

> We begin with a reaffirmation of our commitment to the central beliefs of Friends. . . . Our religious beliefs require that we actively engage in this process. We cannot examine the AFSC without also examining the Society of Friends. . . . The problem is not [our] diversity, but how Friends react to the diversity. . . . The question we Friends, in and out of the AFSC, need continually to ask ourselves is: "Do we believe the Light is present in the other persons and are we in communion with that Light?". . . When as Friends relating to the AFSC we find ourselves on a particular side of a given issue, may we try to respond by committing ourselves to a fuller recognition of the Light in others. . . . The strength of our beliefs, which often leads to diversity, is one of the Society's greatest resources. . . . The Religious Society of Friends is one body with many members. As one member of that body, the AFSC accepts responsibility for engaging in a continuing process of re-examination of its methods, policies and renewal of its witness for truth. . . . One of the dilemmas created by the involvement of the AFSC in issues of great complexity is how in a timely way to reach and involve the broader community of Friends. . . . Ours is an adventure of faith, as is that of the Society of Friends. . . we pray that [AFSC and the Society of Friends] will strengthen each other.[37]

In February 1980 the AFSC sent a letter to all clerks and pastors of monthly meetings concerning the death penalty issue and in May Francis G. Brown, secretary of Philadelphia Yearly Meeting, sent a letter signed by several Quaker groups, calling attention to the one sent by AFSC. This letter noted "the signer's appreciation of AFSC's undertaking consultation with Friends on the death penalty issue."[38]

Also in 1980 both the AFSC and the Yearly Meeting were deeply involved in the question of South Africa's policy of apartheid. What is the right course to follow in the investment of funds? Should funds belonging to the AFSC and to the Yearly Meeting be withdrawn from companies which do business in South Africa? This issue, along with efforts to repeal the draft and the death penalty claim the active attention of both AFSC and the Yearly Meeting.

---
37. Report of the Ad Hoc Committee of AFSC/Friends Concerns, June 26, 1980.
38. Letter from Louis Schneider, June 12, 1980.

The Yearly Meeting and the A.F.S.C.

John A. Sullivan, AFSC staff member, addressed one of the 1980 sessions of the Yearly Meeting on "Facing the 1980's." He indicated that Friends were wise when they revised the latest edition of *Faith and Practice* in 1972 and restrained themselves from producing a blue-print of:

> what the Kingdom of God on earth would be in terms of some ambitious organizational plan. . . . They were unequivocating in what some elements of the Kingdom should include: "a social order free of violence and oppression, self-determination, social and economic justice, adequate food, shelter, education and love for all people in all countries. . . ." That is vision enough for Friends in the 80's. . . . As a Quaker organization, the AFSC espouses such goals. . . . AFSC is a Quaker organization, but it reaches beyond Quakerism.[39]

In the same session, Edwin B. Bronner, AFSC Corporation member, also addressed the Yearly Meeting and said that we are no longer, either in the AFSC or the Yearly Meeting, "going to seek for unity in those points about which we are all agreed or shall we try to find it also in those things wherein we differ. . . . Friends who disagree about specific policies (such as tax refusal, investments, etc.) may at the same time find themselves united in concern. When each is prayerfully seeking light, all may attain to fuller understanding of the right course. . . . Friends should seek in a loving spirit to find their way through the differences to a new unity."[40]

The new Executive Secretary of the AFSC, Asia Alderson Bennett, appointed in April of 1980, became a member of Philadelphia Yearly Meeting (Arch Street) at an early age, and remained in that Yearly Meeting until she and her husband joined Seattle, Washington Meeting. "I want to know more about some of the other branches of Quakerism," she says. "I want the AFSC to benefit from a relationship with the whole Quaker movement. . . . Learning and growing is part of the excitement and the challenge."[41]

---

39. John A. Sullivan, April 2, 1980.
40. Edwin B. Bronner, Statement to the Yearly Meeting, April 2, 1980.
41. *Friends Journal,* Vol. 26, No. 15, Oct. 15, 1980, p. 6.

Appendices
Prepared by BARBARA L. CURTIS

A Chronology of Philadelphia Yearly Meeting

1674   Acquisition of West Jersey by Friends John Fenwick and Edward Billing.

1675   "Concessions and Agreements of the Province of West Jersey" is signed. Quaker colony at Salem is established in Fall. Some Friends cross the Delaware to make their homes at Upland, Pa.

1677   A colony of Quakers settles at Burlington, N.J.

1681   On September 10 (N.S.) a General Meeting of Friends is held at Burlington. This body ultimately became Philadelphia Yearly Meeting.

1682   The ship "Welcome" arrives in Delaware Bay and the first settlement is made at Philadelphia by William Penn and others.

1682   First Friends meeting is held in Delaware at New Ark, near Claymont.

1685   Sessions of Philadelphia Yearly Meeting are held in Philadelphia in September. A meeting of ministers took place on the day before the opening of yearly meeting sessions.

1686   The two sessions of Philadelphia Yearly Meeting, one held in Burlington and one held in Philadelphia are combined in this year. Thereafter in even years the meeting was in Burlington, in odd years in Philadelphia.

Appendices

1687   Formal recognition by Friends of a specific testimony against the sale of liquor to the Indians.

1688   Protest against slavery minuted by the Friends group in Germantown, Pa. and sent forward to Abington Monthly Meeting held at Dublin, Pa., and thence to Philadelphia Quarterly Meeting. That body referred the letter to the Yearly Meeting where it was read, but no action taken.

1689   Founding of the Friends Public School in Philadelphia.

1692   Keithian controversy flares. Philadelphia Yearly Meeting testifies against George Keith in a session held at Burlington, N.J.

1694   Philadelphia Yearly Meeting records an advice regarding the children of Friends marrying out of the Society.

1696   Erection of the Great Meeting House at S.W. corner of 2nd & High Streets, Philadelphia. The house was designed to accommodate the Yearly Meeting as well as Philadelphia Monthly Meeting.

1697   Philadelphia Yearly Meeting sends relief to Friends in New England where crop failures had put Friends in severe need.

1701   William Penn grants the settlers in Pennsylvania a Charter of Privileges.

1702   Quaker and other proprietors of the East and West Jerseys relinquish control of the civil government to the Crown.

1704   Discipline of Philadelphia Yearly Meeting is drawn up, put into writing and approved for use by all quarterly and monthly meetings.

1714   Yearly Meeting sessions at Burlington are so well attended that the courthouse had to be used in addition to the Friends meeting house.

1715   The epistle from Philadelphia Yearly Meeting to the quarterly and monthly meetings exhorts Friends to carry out the advice against importing or trading in slaves.

1718   The meeting of ministers of Philadelphia Yearly Meeting becomes the Meeting of Ministers and Elders.

1719   A revised edition of the Discipline, still in manuscript form only, is prepared and circulated to all meetings.

Friends in the Delaware Valley

1725	The provision for the use of affirmation instead of oath taking, made by the Pennsylvania colonial legislature, becomes finally acceptable to the Crown.
1730	Philadelphia Yearly Meeting makes the purchase of imported slaves a disownable offence.
1735	Philadelphia Yearly Meeting appeals to London for support of the colonial government of Pennsylvania in resisting the claim of Lord Baltimore to the three lower counties of Pennsylvania.
1737	Additional land is acquired from the Indians by the government of Pennsylvania in the "Walking Purchase."
1741	James Logan writes to Philadelphia Yearly Meeting to point out the incompatibility of Friends' peace position and the holding of public office during a period of war.
1743	A group of Queries is prepared by the Yearly Meeting to help the subordinate meetings report on their state.
1746	Philadelphia Yearly Meeting urges monthly meetings to establish schools for both religious and secular education.
1747	The book of discipline is revised once more.
1750	Death of John Kinsey, clerk of the yearly meeting for 20 years. He also served during this time as speaker of the Assembly and as chief justice of the province.
1753	2nd and Pine Street Meetinghouse erected and used by the Yearly Meeting of Men Friends. Yearly Meeting agrees to adopt the Gregorian calendar as a result of notification from London Yearly Meeting that this change was being made worldwide. Quaker numerical nomenclature was retained in stating days and months.
1754	Epistle of Philadelphia Yearly Meeting deplores slavery and the condition of blacks.
1755	Outbreak of French and Indian War. Most Quakers withdraw from the Pennsylvania Assembly.
1756	Philadelphia Yearly Meeting establishes the Meeting for Sufferings. Friendly Association for providing aid to the Indians is founded.
1760	Last of the alternate year sessions of Yearly Meeting is held at Burlington, N.J.

Appendices

1762	Revision of the discipline is completed. The application of the peace testimony and the encouragement of moves toward the abolition of slave holding among Quakers is emphasized.
1764	"Paxton boys" march on Philadelphia. The Yearly Meeting went on record as opposing taking up arms even in a defensive effort.
1772	Philadelphia Monthly Meeting, numbering about 2,000 members, divides into three monthly meetings: Philadelphia Monthly Meeting, Philadelphia Monthly Meeting for the Southern District and Philadelphia Monthly Meeting for the Northern District.
1774	Yearly Meeting directs disownment for any Friend holding slaves or engaged in commerce in slaves.
1776	Yearly Meeting undertakes a relief effort on behalf of the city of Boston and Friends in New England.
1776	Yearly Meeting issues an epistle, "The ancient testimony and principles . . . received with respect to the King and government and touching the commotions now prevailing . . ."
1777–1778	Leading Philadelphia Quakers arrested and sent as exiles to Virginia.
1777	Yearly Meeting advises members not to sell grain to be distilled, or to partake of spirituous beverages.
1781	A dissident yearly meeting, known as the Free Quakers, is started as a protest against Philadelphia Yearly Meeting enjoining all Friends from participation in the Revolutionary War. A meetinghouse for this body was erected at 5th and Arch Sts. in 1783.
1784	Yearly Meeting submits an address to Congress calling for the prohibition of the importation of slaves.
1789	Address of the Quakers to George Washington, assuring him of their loyalty to the new civil government.
1797	First printed Discipline is issued with title "Rules of Discipline and Christian Advices."
1799	Westtown Boarding School is founded by the Yearly Meeting.

Friends in the Delaware Valley

1804   Meetinghouse at 4th and Arch Streets opens with only the East wing complete. West wing was first ready for use in 1811.

1817   Friends Asylum opens under a Quaker board of managers. Tract Association of Friends of Philadelphia is established.

1819   Jonathan Evans, elder in Philadelphia Monthly Meeting, publicly states his disapproval of the doctrines of Elias Hicks.

1827   Supporters of Elias Hicks, mostly country Friends, withdraw to Green Street Meetinghouse under the leadership of John Comly. They declare themselves to be Philadelphia Yearly Meeting. The remaining Friends at Arch Street Meetinghouse also call themselves Philadelphia Yearly Meeting.
Periodical publication of *The Friend* by a group of Orthodox Quakers is begun.

1829   Orthodox Friends establish the Bible Association of Friends of Philadelphia.

1830   "The testimony of the Society of Friends in the continent of North America" is published by Orthodox Friends as a statement of faith for Philadelphia and seven other yearly meetings.

1833   The Haverford School (later, College) opens for the higher education of young men of the Society of Friends (Orthodox).

1834   Publication by Hicksite Friends of *Friends Miscellany* and the eight volume set *The Works of George Fox*.

1837   Orthodox Friends begin the publication of *Friends Library*, a 14 volume set of some of the writings of early Quakers.

1838   Hicksite Friends begin publication of the periodical *Friends Intelligencer.*

1838   Burning of Pennsylvania Hall in Philadelphia where John G. Whittier and Lucretia Mott and others had spoken against the evils of slavery.

Appendices

1843	Census of the black community in Philadelphia is conducted by Friends.
1847	Gurneyite Friends start their own publication *Friends Review*.
1853	Progressive or Congregational Friends meet at Longwood, Pa.
1857	Philadelphia Yearly Meeting (O.) discontinues correspondence with other yearly meetings.
1860	Gurneyite Friends establish Friends First Day School Association.
1861	Philadelphia Yearly Meeting (H.) accords full recognition to the Meeting of Women Friends in the affairs of the yearly meeting.
1863	Orthodox Friends start the Friends' Freedmen's Association. Two years later the Hicksite body formed the Association for the Aid and Elevation of the Freedmen.
1864	Swarthmore College corporation is formed by a group of Philadelphia and other Hicksite Quakers. The college opens to students in 1869.
1866	Philadelphia Yearly Meeting (H.) records approval of efforts to seek the unity of all differing kinds of Quakers.
1867	Peace Association of Friends in America is organized by Friends (O.) outside of Philadelphia. Individual Philadelphia Quakers take part.
1869	Philadelphia Yearly Meeting (O.) joins other yearly meetings in forming The Associated Executive Committee of Friends on Indian Affairs.
1872	Representative Meeting (H.) petitions the Pennsylvania legislature to strictly control the liquor trade.
1874	Philadelphia Yearly Meeting (H.) admits women as members of Representative Meeting.
1881	Philadelphia Yearly Meeting (H.) hires a consultant in education to advise its schools.
1883	Gurneyite women Friends begin a mission effort in Japan. This is not officially supported by either of the yearly meetings at first.

Friends in the Delaware Valley

1884	Bryn Mawr College opens under the sponsorship of Orthodox Friends in Philadelphia and Baltimore to provide for the higher education of women.
1891	The Committee on Philanthropic Labor is established by Philadelphia Yearly Meeting (H.)
1893	Gurneyite publication *Friends Review* is merged with the *Christian Worker,* under the editorship of Rufus M. Jones, to become *The American Friend.*
1894	George School, a boarding school for the children of Friends and others, is founded by Philadelphia Yearly Meeting (H.)
	A revised discipline is published by Philadelphia Yearly Meeting (H.)
1897	A committee on Peace and Arbitration is established by Philadelphia Yearly Meeting (H.)
	Philadelphia Yearly Meeting (O.) issues an epistle "To all meetings bearing the name of Friends."
1900	Friends General Conference is formally established. The largest body in the conference is Philadelphia Yearly Meeting (H.)
	Philadelphia Yearly Meeting (O.) establishes a Committee for aid to the Doukhobors.
1901	A peace conference of all American Friends is held in Philadelphia under the auspices of both yearly meetings.
1903	The Education Committee (O.) decides to call its consultant a Visiting Superintendent.
1906	The epistle of London Yearly Meeting to Philadelphia Yearly Meeting (O.) is acknowledged by the clerk.
1910	A General Nominating Committee is created by Philadelphia Yearly Meeting (H.) to provide membership for its many committees.
	Philadelphia Yearly Meeting (O.) revises its discipline.
1911	Friends Central Bureau opens under the auspices of Philadelphia Yearly Meeting (H.) to provide secretarial and other services for the committees.
	Philadelphia Yearly Meeting (O.) changes the name of Meeting for Sufferings to Representative Meeting.

Appendices

1914	Philadelphia Yearly Meeting (O.) develops informal relationship with Five Years Meeting organization. Both Philadelphia bodies protest to President Wilson concerning the use of United States troops in Mexico. Philadelphia Yearly Meeting (O.) agrees to the service of women on its Representative Meeting.
1915	Philadelphia Yearly Meeting (H.) establishes a Committee on Peace and Emergency Service. Peace conference of Young Friends, held at Winona Lake, Ind., is attended by Quakers of all American yearly meetings. The Advancement Committee of Friends General Conference opens Woolman House. This later became Woolman School on the Swarthmore College campus.
1916	Philadelphia Yearly Meeting (O.) sends its first epistle in over 70 years to meetings in all branches of the Society of Friends.
1917	The American Friends Service Committee is founded. Philadelphia Yearly Meeting (O.) creates a Social Order Committee.
1918	Philadelphia Yearly Meeting (O.) sets up the office of Secretary of Representative Meeting, which was transformed in 1920 into Secretary of the Yearly Meeting.
1920	Both Philadelphia bodies take official part in the All-Friends Conference held in London. Young Friends of both Philadelphia yearly meetings participate in a national Young Friends conference at Richmond, Ind.
1921	A letter concerning peace is prepared by Friends from both yearly meetings working jointly. Philadelphia Yearly Meeting (O.) establishes the Mission Board which combined the work of its Committee on Christian Labor in Foreign Lands with the older independent Foreign Mission Association of Friends.
1923	Friends Historical Society of Philadelphia and Friends Historical Association merge into one organization called Friends Historical Association.

Friends in the Delaware Valley

1924   In Philadelphia Yearly Meeting (H.) the separate Men's and Women's Meetings are united in one yearly meeting under the clerkship of Jane P. Rushmore.
Friends Social Union is formed, for men only at first, to help the two yearly meetings grow toward unity.

1929   Committee on the Interests of the Colored Race of Philadelphia Yearly Meeting (H.) changes its name to Race Relations Committee and begins to function jointly with the committee of the same name in Philadelphia Yearly Meeting (O.)
Pendle Hill at Wallingford is purchased and prepares to open as a center for Quaker and community studies.
Philadelphia Yearly Meeting (O.) unites the Men's and Women's Meetings into one body.

1930   Young Friends groups in both yearly meetings join together to form the Young Friends Movement.

1932   Temperance Association receives appointments from both Yearly Meetings and thus becomes united.

1933   Chestnut Hill Friends Meeting is launched as a United Meeting, with membership in both yearly meetings.
Formation of a joint Friends Peace Committee with office at 304 Arch St., and library at 1515 Cherry St.

1936   Philadelphia Yearly Meeting (O.) appoints a committee from Ministers and Elders to cooperate with the equivalent group from Philadelphia Yearly Meeting (H.)

1938   Philadelphia Yearly Meeting (O.) appoints members to serve on Social Service Committee of Philadelphia Yearly Meeting (H.) In 1941 this joint service was abandoned because of differences in purpose and organization of the two separate committees.

1940   Joint session of the two Philadelphia yearly meetings, of one day only, to discuss peace issues.

1943   Marriage Council of the two bodies is functioning as a joint undertaking.

1945   The first black student is enrolled at Westtown. George School admitted its first Negro student in the year following.

Appendices

1946	Philadelphia General Meeting is created by the two Quaker bodies and meets under the clerkship of M. Albert Linton with Eleanor Stabler Clarke serving as recording clerk.
1950	A statement of the Social Order Committee of Philadelphia Yearly Meeting (O.) called "Peace and Social Justice" is endorsed by both yearly meetings.
1954	After a one year trial a *Book of Church Government* is approved by both bodies. This was to be part of a new jointly approved Discipline for the anticipated reunited Philadelphia Yearly Meeting.
1955	The two yearly meetings are united in a session held on Third month 28th and become once again Philadelphia Yearly Meeting of the Religious Society of Friends. Joint presiding clerks were Charles J. Darlington and James F. Walker.

Friends in the Delaware Valley

Clerks—Philadelphia Yearly Meeting—to 1827

Men's Meeting*

1696–1701 Phineas Pemberton
1702 Griffith Owen
1704 Caleb Pusey & Anthony Morris
1710 Anthony Morris
1711–1729 Isaac Norris
1730–1749 John Kinsey
1750–1759 Israel Pemberton, Jr.
1760 John Smith
1761–1766 James Pemberton
1767 George Churchman
1768–1776 James Pemberton
1777 Isaac Jackson
1778–1781 James Pemberton
1782–1786 John Drinker
1787–1788 Caleb Carmalt
1789–1794 Nicholas Waln
1795–1807 Jonathan Evans
1808 John Cox
1809–1810 Jonathan Evans
1811–1816 John Cox
1817–1827 Samuel Bettle

Women's Meeting**

1726 Hannah Hill
1728 Ann Stevenson
1729–1743 Grace Lloyd
1744–1759 Mary Jordan
1760–1777 Mary Pemberton
1778–1794 Hannah Catherall
1795 Mary Pleasants
1796 Mary Gibbons
1797–1798 Hannah Evans
1799 Anne Mifflin
1800–1805 Mary Taylor
1806–1818 Catherine Morris
1819 Ruth Ely
1820–1824 Hannah Lewis
1825 Ruth Ely
1826–1827 Hannah Paul

*Before 1711 the minutes do not clearly indicate a clerk. The names given prior to this date appear to have functioned in that capacity at the dates given.

**Established in 1684. No clerk's name appears until 1726.

Clerks—Philadelphia Yearly Meeting (Orthodox)—1827–1955

Men's Meeting

1827–1830 Samuel Bettle
1831–1861 William Evans
1862–1864 Joel Evans
1865–1876 Joseph Scattergood
1877–1880 Clarkson Shepperd
1881–1896 Joseph Walton
1897–1901 Ephraim Smith

Women's Meeting

1827–1832 Hannah Paul
1833–1834 Ruth Ely
1835–1841 Hannah Rhoads
1842 Beulah H. Nicholson
1843–1845 Hannah Rhoads
1846 Beulah H. Nicholson
1847–1850 Hannah Rhoads

Appendices

Clerks—Philadelphia Yearly Meeting (Orthodox)—1827-1955

Men's Meeting
1902-1905 William Evans
1906-1907 William Bishop
1908-1911 Charles Carter
1912-1923 Davis H. Forsythe
1924-1928 John D. Carter

Joint Clerkship
1929-1939 D. Robert Yarnall
1940 Harold Evans
1941-1943 William Wistar Comfort
1944-1949 Harold Evans
1950-1955 James F. Walker

Women's Meeting
1851 Beulah H. Nicholson
1852-1858 Hannah Rhoads
1859-1866 Elizabeth Pierson
1867-1868 Hannah Warner
1869-1870 Susanna F. Sharpless
1871-1872 Jane Gibbons
1873 Susanna F. Sharpless
1874-1877 Jane Gibbons
1878-1885 Susanna F. Sharpless
1886-1893 Hannah Evans
1894-1896 Elizabeth Smedley
1897-1906 Anna P. Haines
1907-1913 Rebecca Conard
1914-1919 Jane Bartlett
1920-1921 Anna Rhoads Ladd
1922 Mary R. Williams
1923-1927 Anna Rhoads Ladd
1928 Elizabeth B. Jones

Clerks—Philadelphia Yearly Meeting (Hicksite)—1827-1955

Men's Meeting
1827-1829 Benjamin Ferris
1830-1833 John Comly
1834-1836 Joseph Parrish
1837-1842 Benjamin Price, Jr.
1843-1853 James Martin
1854-1864 William Griscom
1865-1872 Dillwyn Parrish
1873-1885 Benjamin G. Foulke
1886-1900 Emmor Roberts
1901-1903 Isaac H. Hillborn
1904-1912 Wesley Haldeman
1913-1917 Joseph T. Foulke

Women's Meeting
1827-1829 Rebecca Comly
1830 Lucretia Mott
1831 Rebecca Comly
1832 Deborah F. Wharton
1833-1836 Lucretia Mott
1837-1844 Deborah F. Wharton
1845-1866 Mary Lippincott
1867-1869 Elizabeth Eastburn
1870-1875 Phebe Foulke
1876-1879 Martha Mellor
1880-1882 Phebe Foulke
1883 Mary Barnard

Friends in the Delaware Valley

Clerks—Philadelphia Yearly Meeting (Hicksite)—1827–1955

Men's Meeting
1918–1922 Morgan Bunting
1923 Walter H. Jenkins

Women's Meeting
1884–1892 Margaretta Walton
1893–1921 Sarah Griscom
1922–1924 Jane P. Rushmore

Joint Clerkship
1924–1925 Jane P. Rushmore
1926–1928 George A. Walton
1929 Jane P. Rushmore
1930–1932 George A. Walton
1933–1941 Thomas A. Foulke
1942–1949 Gordon P. Jones
1950–1953 William Eves III
1954 Charles J. Darlington

Clerks—Philadelphia Yearly Meeting (Reunited)—after 1955

1955 Charles J. Darlington and James F. Walker (Joint Clerks)
1956–1961 Charles J. Darlington
1962–1965 David G. Paul
1966–1967 Albert B. Maris
1968–1972 Charles K. Brown III
1973–1975 Allen J. White
1976–1978 Barbara S. Jacobson
1979– Thomas S. Brown

Appendices

Sites of Sessions of Philadelphia Yearly Meeting, 1681-1981

1681   At home of Thomas Gardner, Burlington, New Jersey.

1683   Yearly Meeting sessions were held at Burlington and then a month later in Philadelphia.

1685   Philadelphia Yearly Meeting met in Philadelphia at the Bank Meetinghouse in the 200 block north and on the west side of Front Street. The house was erected in 1683.

1686   And in alternate years thereafter until 1760, sessions were held in Burlington in the hexagonal meetinghouse on High Street until 1716. From that date on they were held in the larger meetinghouse around the corner on Broad Street, near High Street.

1697   The annual sessions in odd years were held in Philadelphia in the Great Meetinghouse at 2nd and High (Market) Streets built in 1696.

1714   Courthouse at Burlington used for some sessions of Philadelphia Yearly Meeting.

1753   Meetinghouse, called "The Hill Meeting" at 2nd and Pine Streets, Philadelphia, completed in time for use for sessions of Yearly Meeting.

1755   The Greater Meetinghouse, on the site of the older Great Meetinghouse, used for the first time for sessions of Yearly Meeting. The location was 2nd and High (Market) Streets as before.

1805   A portion of the meetinghouse complex at 4th and Mulberry (Arch) Streets was completed and used by the Women's Yearly Meeting. This was the East Meeting Room.

1811   Keys Alley Meetinghouse, near 2nd and Race Streets, erected in 1789, was being used in 1811 for sessions of Men's Yearly Meeting. They adjourned from that location to the large meetinghouse complex at 4th and Mulberry Streets. The men met in the East Room and the women in the recently completed West Room.

1828   Cherry Street Meetinghouse, built by the "separated" Friends near 4th and Cherry Streets in time for the Yearly Meeting of Philadelphia Friends (Hicksite) was used by women's meeting. Men held their sessions in the meetinghouse at 4th and Green Streets.

1856   Race Street Meetinghouse, located on Race Street west of 15th Street, was built when the Cherry Street House could no longer adequately serve the Yearly Meeting (Hicksite). This new house was first used for Yearly Meeting sessions in 1857.

1955   Meetinghouse at 4th and Arch Streets is now used by the reunited Philadelphia Yearly Meeting for sessions of the annual gathering. The Race Street complex, now called the Friends Center, houses the offices of the Yearly Meeting and other Friends activities.

Index

Abington Monthly Meeting, 189, 249
Abington Quarterly Meeting, 59, 75, 109
Abolition, see Slavery
Abstinence, see Temperance
Advices, 16, 27, 127. See also Queries
Affirmations, see Oaths and Affirmations
Albertson, Henry, 150
All Friends Conferences, see Friends World Conferences
Allen, William C., 131
American Friend, 111, 121, 131, 254
American Friends Fellowship Council, 164
American Friends Service Committee, 6, 147, 150, 164, 167, 197, 217, 225, 226, 227, 230, 235-247, 255
American Red Cross, 237
"Ancient Testimony," 27, 28, 47
Arch Street Meeting House, 59, 69, 78, 79, 148, 181, 252
Arch Street Monthly Meeting, 83
Austin, Ann, 174

Bacon, Margaret, 5
Baily, Joshua L., 121
Balderston, Lloyd, 217, 222

Bank Street Meeting House, 180
Barclay, Robert, 3, 19, 80
Barnard, Mary, 259
Bartlett, Jane, 259
Bean, Hannah and Joel, 132
Benezet, Anthony, 31, 37, 40, 44, 51, 96
Benjamin, Philip S., *Philadelphia Quakers in the Industrial Age,* 112, 116, 118, 120, 126, 142-143, 150
Bennett, Asia Alderson, 247
Berean, The, 72, 77
Bettle, Samuel, 76, 97, 258
Bible Association (1829), 81, 252
Biddle, Clement, 50
Billing, see Byllynge
Binford, Elizabeth and Gurney, 220, 222, 225
Bishop, William, 259
Black Economic Development Conference, 13
Blacks, 106, 117-19, 157-59, 187, 250, 256. See also "Contrabands," Freedmen, Slavery
Bond, Elizabeth Powell, 129, 240
Book of Church Government (1954), 257. See also Discipline
Book Store, Arch Street, 142
Borton, C. Walter, 223, 224
Boudinot, Elias, 50

Index

Bowles, Gilbert, 145, 218–19, 221, 223, 224, 225, 227
Bowles, Herbert, 218
Bowles, Luanna, 222, 227
Bowles, Minnie Pickett, 223, 224, 225
"Box Meeting," 174
Braithwaite, Anna, 75, 184, 186
Braithwaite, George Burnham, 219
Brinton, Anna Cox, 169, 230, 231
Brinton, Edward, 133
Brinton, Howard, 120, 133, 223, 230
Brit, Thomas, 16
British Friends, 20, 26, 39, 40, 41
Brock, Peter, 112, 123
Bronner, Edwin B., 4, 247
Brown, Charles K. III, 260
Brown, Ernest L., 236
Brown, Francis G., 246
Brown, Thomas S., 260
Bryan, Helen R., 241
Bryn Mawr College, 128, 129, 135, 254
Buck Hill Falls Inn, 133
Bucks Quarterly Meeting, 16, 44, 75
Bunting, Morgan, 260
Burlington and Bucks Quarterly Meeting, 106
Burlington Monthly Meeting, 15, 78
Byllynge, Edward, 15, 248

Cadbury, Henry J., 134, 141, 147, 162, 235
Cadbury, Dr. William, 140, 231
Caln Quarterly Meeting, 192
Carmalt, Caleb, 258
Carpenter, Samuel, 32
Carter, Arabella, 235
Carter, Charles, 259
Carter, John D., 259
Cary, Eleanor, 236
Catawissa Indulged Meeting, 25
Catherall, Hannah, 258
Cecil Monthly Meeting, 59

Central Bureau, 142, 197
Central Philadelphia Monthly Meeting, 245
Centre Meeting, 59
Charming Polly, 46
Charter of Privileges, (1701), 23, 249
Chester Monthly Meeting, 202
Chester Quarterly Meeting, 16, 17, 36
Chesterfield Monthly Meeting, 15
Chestnut Hill Monthly Meeting, 165, 256
Cheyney State College, 119
Chichibu, Princess, 221
Child, Isaac, 180
China, 140, 156, 223
Christian Labor in Foreign Lands, Committee on, 156, 216–17
Christian Quakers, 20
Christian Worker, 111, 254
Church of Christ in Japan, 224, 225
Church World Service, 230
Churchman, John, 37, 38, 40, 44, 180
Churchman, Margaret, 180
Civil Liberties, 153–54
Civil War, 112–16, 194
Civilian Public Service, 244
Clarke, Eleanor Stabler, 168, 257
Clevenger, Janice, 228
Clothier, Isaac H., 133
Coale, Edith, 236
Coale, Josiah, 14
Coates, Beulah, 178
Cocks, William H., 235
Coffin, Levi, 117
Collier, John, 212–13
Comfort, Henry W., 235
Comfort, William Wistar, 259
Comly, John, 76, 191, 252, 259
Comly, Rebecca, 259
Comstock, Elizabeth, 132–33, 189
Conard, Rebecca, 259
Concord Monthly Meeting, 202

Concord Quarterly Meeting, 97, 153
"Contraband," 116-17, 187
Cook, Charity, 182
Cope, Thomas Pym, 61, 83
Cornplanter, 205
Cosand, Joseph and Sarah, 215, 218
Cox, John, 258
Crosswicks Meeting, 59

Daimachi Center, Tokyo, 218, 225, 227, 228
Darby Monthly Meeting, 59
Darlington, Charles, 169, 257, 260
Discipline, 9, 15-16, 24, 25, 27, 28 34, 47, 80, 84, 134, 167, 169, 249, 250, 251, 254. See also *Faith and Practice*
Disownment, 20, 28, 50, 78, 80, 84, 126-27, 134, 251; in Civil War, 114-115
Divorce, 159
"Doctrine of Christianity," 36
Doherty, Robert, 67
Doukhobors, 109, 121-23, 188, 254
Dow, Neal, 99
Downing, George V., 236
Drinker, Elizabeth, 182
Drinker, Henry, 45, 48, 52, 57, 58, 63
Drinker, John, 258
Dublin Monthly Meeting (Abington), 18
Duck Creek Meeting, 59
Duguid, Mary and William, 236
Dungan, Margaret, 155

East Jersey, 14, 15, 19
Ecumenical movement, 8
Education, 51, 86-91, 100
Education Committee (Hicksite) 110-11, 161, 192, 253 (Orthodox) 109-110, 117, 161, 254. See also Friends Council on Education
Eighteenth Amendment, see Prohibition

Elders, see Ministers and Elders
Elkinton, Howard and Katharine, 236
Elkinton, J. Passmore, 222
Elkinton, Joseph S., 122, 139, 190
Elkinton, Mary, 216
Elkinton, Sarah, 139
Ely, Ruth, 258
Estaugh, Elizabeth, 181
Estaugh, John, 181
Evangelicals, 4, 66-71, 76, 79, 81, 185
Evans, Harold, 259
Evans, Joel, 258
Evans, Jonathan, 74, 86, 252, 258
Evans, William, 103, 258, 259
Eves, William III, 149, 163, 260
Exeter Monthly Meeting, 25

Fair Hill Meeting House, 140
Faith and Practice, 171-172, 242
Fell, Margaret, 5-6, 174, 175
Fell, Sarah, 175
Fellowship Committee, 244
Fellowship Council, 244
Fenwick, John, 15, 248
Ferris, Benjamin, 75, 259
Ferris, David, 72
Ferris, Henry, 140, 150
Fifth and Cherry Street Meeting House, 79, 193
First Day Schools, 11, 106, 108, 125-26, 142
Fisher, Samuel Logan, 60
Fisher, William Logan, 85
Five Years Meeting, 130, 131, 231, 255
Foreign Missionary Association, 140, 215-217, *passim*
Forster, William, 75
Forsythe, Davis H., 259
Fothergill, Samuel, 38
Foulke, Benjamin, 104, 239
Foulke, Eliza Ambler, 163, 171, 229, 231
Foulke, Joseph T., 259

Index

Foulke, Phebe, 259
Foulke, Thomas, 171, 229, 260
Fourth and Arch Street Meeting House, see Arch Street Meeting House
Fox, George, 3, 14, 65, 80, 84, 85, 173, 174, 200-01; *Journal* of, 223, *Works,* 252
Free Produce movement, 93, 95
Free Quakers, 50, 251
Freedmen, Friends work for the, 116-117, 194, 253
Friend, The, 77, 80, 83, 90, 103, 110, 113, 121, 123, 131, 140, 165-66, 252
Friend: Advocate of Truth, 77
Friendly Association, 7, 203-05, 250
Friends Asylum, 61, 62, 78, 252
Friends Center, 244-45
Friends Center in Tokyo, 224, 226, 229, 230, 232
Friends Central Bureau, 108, 254
Friends Central School, 90
Friends Committee on National Legislation, 165
Friends Council on Education, 161, 232
Friends First Day School Association, 160, 253
Friends Food Unit, 146
Friends Freedmen's Association, 116, 187, 253
Friends General Conference, 108, 130, 135, 140, 141, 145, 197, 245-46, 254
Friends Girls School, Tokyo, 7, 156, 215, 219, 220, 221, 223, 224, 225, 226, 230, 232, 233
Friends Historical Association, 133, 164, 255
Friends Historical Society, 164, 255
Friends Institute, 236
Friends Institute, Tokyo, 219
Friends Intelligencer, 111, 113, 124, 134, 139, 140, 141, 143, 149, 153, 161, 162, 163, 165-66, 252

Friends Journal, 111, 134, 166
Friends Library: Comprising Journals, etc., 79-80, 252
Friends Miscellany, 80, 252
Friends National Peace Conference, 143, 235
Friends National Service Committee, see American Friends Service Committee
Friends Neighborhood Guild, 125-26
Friends Public School, see William Penn Charter School
Friends Quarterly, 133
Friends Review, 83, 111, 113, 129, 130, 253, 254
Friends Select School, 241
Friends Service Council (British), 237
Friends Social Union, 164, 256
Friends Tract Association, 63, 252
Friends War Victims Relief Committee (British), 237
Friends World Committee for Consultation, 9, 164, 167, 244
Friends World Conferences: London, (1920), 9, 147-48, 255; Swarthmore and Haverford, 164, 222-23; Oxford, (1952), 9, 242; Greensboro, N.C., 9
Frost, J. William, 3, 11-12
Furnas, Elizabeth Walters, 163

Gardner, Thomas, 15, 176, 261
Garrett, John B., 209
Garrett, Thomas, 85, 95
Garrison, William Lloyd, 95
General Meetings, 8, 15, 16, 39, See also Philadelphia General Meeting, 1946-1954
Genessee Yearly Meeting, 197
George, John Malin, 107, 196
George School, 107, 111, 130, 135, 159, 192, 196, 254
Germantown Friends Protest against Slavery, 18

266

Germantown Monthly Meeting, 222
Gibbons, James, 72
Gibbons, Jane, 259
Gibbons, Mary, 258
Gilpin, Thomas, 61
Grant, Ulysses S., 118, 207-08
Gray, Isaac, 50
Gray, Violet Gordon, 229
Great Earthquake, (1923), 217
Great Meeting House, 180, 249
Greater Meeting House, 180-81
Green Street Meeting House, 193, 252
Green Street Monthly Meeting, 74, 75, 76, 79
Griscom, Sarah, 260
Griscom, William, 259
Grubb, Sarah, 81
Gurney, Eliza P., 113-14
Gurney, Joseph John, 81, 82, 113, 186
Gurneyite Friends, 5, 82, 83, 109, 111, 112, 117, 118, 126, 128, 130, 131, 132, 135, 160, 186, 253
Gwynedd Monthly Meeting, 59

Haddon, Elizabeth, see Estaugh, Elizabeth
Haddonfield and Salem Quarterly Meeting, 107
Haddonfield Monthly Meeting, 78
Haddonfield Quarterly Meeting, 89
Hadley, Herbert, 8
Haines, Anna P., 259
Haines, Mary H., 222
Haldeman, Wesley, 259
Hancock, Cornelia, 117, 194
Harding, Warren G., 154-55
Harris, Elizabeth, 194
Hartman, Neil and Venette, 229
Hartshorne, Henry, 130
Harvey, William B., 218
Haverford College, 83, 90, 111, 112, 128, 130, 135, 164, 236, 252

Haverford School, 61, 90, 252
Heald, Dr. Pusey, 139
Hepburn, John, 30
Heston, Zebulon, 204
Hicks, Edward, 85, 86
Hicks, Elias, 71, 73-75, 81, 93, 128, 133, 252
Hicksite Friends, 4, 5, 71-85, *passim* 88, 98, 99, 103-121, *passim*, 123-136, 207, 211, 258, 259, 260
Hicksite-Orthodox Separation, 3-5, 12, 63-79, 100, 101, 133-36, 164, 172
Hicksite Women's Yearly Meeting, 185, 191-198
Hill, Hannah, 258
Hillborn, Isaac H. 159
Hirakawa, Seiju, 220, 221, 222
Hoag, Enoch, 209
Holmes, Jesse H., 129, 130, 147, 235
Hooten, Elizabeth, 174
Hoover, Herbert, 212
Hopewell Monthly Meeting, 25
Hopper, Isaac, 93, 95
Hubben, William, 166
Hull, Hannah Clothier, 141, 197
Hull, William I., 124, 147
Hunt, John, 48, 181
Huston, Dr. Charles, 124
Hutton, Addison, 128, 133
Hutton, Susannah, 182

Ibaraki Ken, Friends work in, 218, 219, 223, 224, 230
Indian Association, see Friendly Association
Indian Committees, 7, 83, 118, 160, 188, 192, 196, 200-13, 253
Indian Rights Association, 211
Indians, 6-7, 17-18, 42, 43, 100, 118, 184, 200-14, 249
Institute for Colored Youth, see Cheyney State College

Index

Institute of Race Relations, 241–42
Institutes of International Relations, 239
International Outreach Committee, 233
Interracial Section of AFSC, 239–242, *passim*
Iwahashi, Takeo, 224, 228

Jackson, Halliday, 206
Jackson, Isaac, 258
Jacob, Caroline N., 169
Jacobson, Barbara S., 260
Janney, Samuel M., 209
Japan, Friends Mission in, 7, 126, 140, 156, 171, 188, 215–233, 253
Japan Committee, 156, 230, 231–233
Japan International Christian University, 233
Japan Service Committee, 217
Japan Yearly Meeting, 7, 220, 221, 222, 224, 225, 228, 231, 232
Japanese-American Fellowship, 224
Jeanes, Anna T., 108, 117, 152
Jenkins, Charles F., 146, 167
Jenkins, Howard M., 111, 124
Jenkins, Walter H., 146, 259
Johnson, Andrew, 116
Johnson, Charles S., 241
Johnson Immigration Bill, 217–18
Joint Japan Committee, see Japan Committee
Joint Peace Committee, see Peace Committee
Jones, Eli and Sybil, 215
Jones, Elizabeth B., 131, 259
Jones, Esther Balderston, 217, 218, 219, 221
Jones, Gordon P., 260
Jones, Mary Hoxie, 7
Jones, Owen, Jr., 48
Jones, Rebecca, 180, 181
Jones, Rufus M., 12, 111, 121, 124, 131, 144, 147, 223, 234, 236, 254

Jones, Thomas E., 218, 219, 221
Jordan, Mary, 258

Keith, George, 19–21, 249
Kelley, Abby, 199
Kennett Monthly Meeting, 85
Kerlin, Robert T., 153
Key's Alley Meeting House, 181, 261
Kinsey, John, 32, 35, 37, 250, 258
Kurama, Kikue, 222

Ladd, Anna Rhoads, 259
Lake Mohonk Conference on the Indians, 212
Lake Mohonk Conference on International Arbitration, 124
Lakey, George, 240
Lamb, Edith, 219
Lansdowne Monthly Meeting, 222
LARA (Licensed Agencies for Relief in Asia), 227
Lay, Benjamin, 31
Laymen's Committee Mission, 220–21
Leeds, Morris E., 150, 236
Lewis, Hannah, 258
Lewis, Lucy Biddle, 235
Lewis, Mordecai, 60
Liberals, 72–74
Lightfoot, Susanna, 180
Lincoln, Abraham, 114, 115, 116
Lindley, Laurence, 209
Linton, M. Albert, 150, 163, 168, 259
Lippincott, Horace Mather, 134, 141, 146
Lippincott, J. B., 60
Lippincott, Mary, 259
Liquor, 18, 27–28, 51–52. See also Temperance
Lloyd, David, 23, 32
Lloyd, Grace, 258
Logan, James, 35, 37, 250

London Grove Meeting House, 59
London Yearly Meeting, 16, 19, 20, 26, 30, 44, 79, 81, 130, 132, 135, 138, 150, 155, 175, 181, 185, 254
Longwood Meeting, see Progressive Friends
Lovering, Joseph, 60
Loyalty tests, 48, 49
Lundy, Benjamin, 95
Lynching, Committee to Protest Against, 157

Malara Girls Secondary School, 233
Maris, Albert B., 260
Maris, Robert H., 222
Marlborough Monthly Meeting, 85
Marriage, 16, 84, 159, 194
Marriage Council, 159-60, 256
"Marrying out," 22, 28, 38, 127, 188, 249
Marshall, Clara, 133
Martin, James, 259
Maule, Joshua, 115
Meeting for Sufferings, see Sufferings, Meeting for, Representative Meeting
Meeting schools, 29-30, 88
Mellor, Martha, 259
Menallen Indulged Meeting, 25, 59
Mennonites, 203
Men's meetings, 15, 17, 180-81, 250, 256
Mexico, 109, 144, 189, 215, 255
Michener, Ezra, 113
Middletown Monthly Meeting, 202
Mifflin, Anne, 258
Miho, Fumie, 232
Milhouse, Thomas, 180
Military service (1756), 42-43
Ministers and Elders, 9, 17, 19, 22, 25-26, 38, 39, 58, 69, 76, 160, 256
Ministry and Counsel, 160
Ministry and Worship, Meeting on, 160
Mission Board, 156, 214-33, *passim*, 255. See also Japan Committee
Mito, 219, 220, 228, 230, 233
Moon, James, 190
Moore, Ann, 180
Morris, Anthony, 258
Morris, Catherine, 258
Morris, Homer L., 235
Morris, Mary H., 215, 222
Mosheim, Johan, 71
Mott, James, 85, 194, 195
Mott, Lucretia, 6, 61, 85, 99, 128, 185, 191, 192-93, 194, 195-96, 252, 259
Mt. Holly Meeting, 75

National Council of Churches, 8
Naylor, James, 136
New Ark, Delaware, 248
New Jersey Association for Helping the Indians, 204
Newlin, Edith, 219
Nicholson, Anna Margaret, 230, 232
Nicholson, Beulah H., 258, 259
Nicholson, Herbert and Madeline, 217, 220, 222, 223, 225, 230
Nicholson, Samuel, 230, 232
Nicholson, Sarah, 182
Nicholson, Vincent D., 235, 236
Nicholson, Dr. William, 209
Nitobe, Dr. Inazo, 216
Nitobe Lecture, 221
Norris, Isaac, Sr., 32, 33, 177, 258
Norris, Isaac II, 37
Northern Superintendency, 209
Nutt, Mary, 207

Oaths and affirmations, 33-34
Ohio Yearly Meeting, 60, 82, 97
Oriental Exclusion Act, 217-18
Orthodox Friends, 4, 5, 8, 9, 63-68, *passim*, 88, 98-99, 103-136, *passim*, 208, 209, 258, 259
Orthodox Women's Yearly Meeting, 185-191

Index

Osaka Monthly Meeting, 224, 228
Owen, Griffith, 258

Pancoast, Henry S., 211
Parish, John, 204
Parker, Ely S., 207
Parrish, Dillwyn, 259
Parrish, Joseph, 259
Patterson, Mary T. Sullivan, 163
Paul, David G., 260
Paul, Hannah, 258
Paxton Boys, 42, 251
Peace and Arbitration Committee, 254
"Peace and Social Justice," 257
Peace Association of Friends of America, 124, 253
"Peace Automobiles," 144
Peace Caravans, 239-40
Peace Committees, 144, 154, 155, 158, 236, 238-39, 240, 256
Peace Conference of 1901, 124, 254
Peace Section, AFSC, 239
Peace Testimony, 34-36, 39-43, 46, 50, 123-25, 143-44, 196, 251
Pearson, Alice Lewis, 225
Pelham Monthly Meeting, Ontario, 183-84
Pemberton, Israel, Sr., 32
Pemberton, Israel, 37-38, 40, 46, 48, 203, 258
Pemberton, James, 39, 42, 45, 48, 52, 57, 58, 63, 258
Pemberton, John, 38, 48
Pemberton, Mary, 258
Pemberton, Phineas, 14, 32, 258
Pendle Hill, 141, 164, 222, 230, 232, 256
Penn, Thomas, 203
Penn, William, 3, 6, 15, 16, 17, 20, 23, 24, 176, 201, 248
Pennsylvania Abolition Society, 57, 61, 92
Pennsylvania Hospital, 61
Perry, Edith and Henry, 229

Philadelphia General Meeting (1946-1954), 168, 172, 231, 244, 251
Philadelphia Indian Aid Association, 209
Philadelphia Monthly Meeting, 19, 59, 74, 251
Philadelphia Quarterly Meeting, 16, 21, 39, 43, 57, 75, 84, 105, 249
Philadelphia Yearly Meeting, Joint Session (1945), 226
Philadelphia Yearly Meeting of Women, see Women's Meetings
Philanthropic Labor, Committee on, 107, 142, 152, 196, 197, 240, 242, 254
Phillips, Caroline, 163
Pickett, Clarence E., 169, 240, 244-45
Pierson, Elizabeth, 259
Pine Street Meeting House, 181, 250
Platt, Joseph, 140
"Plea for Unity," 169
Pleasants, Mary, 258
Pocono Lake Improvement Ass'n, 148
Pocono Manor, 134
Post, Christian Frederick, 204
Potts, Sarah, 222
Potts, Thomas C., 222
Preparative meetings, 10
Preston, Ann, 183
Price, Benjamin, Jr., 259
Primitive Friends, 4
Prison Committee, 159
Progressive Friends, 4, 85, 99, 253
Providence Monthly Meeting, 59
Publications, 18, 62
Pusey, Caleb, 258

Quarterly Meetings, 10-11, 16, 17
Queries, 27, 38, 45, 109, 127, 179, 195, 250
Quietism and Quietists, 65, 66, 67, 68, 71, 81, 83, 84, 110

Race Relations committees, 158, 190, 240, 241, 242, 256
Race Street Friends, see Hicksite Friends
Race Street Meeting House, 79, 193
Radnor Monthly Meeting, 65, 195
Ramallah, 215
Reconciliation, Letter of (1916), 138-39
Reform movement, 37-39, 50-51
Religious Education Committee, 160-61
Representative Meetings, 84, 109, 110, 142, 189-90, 195, 233, 245, 254, 255
Reynell, John, 40, 45, 46
Rhoads, Charles James, 212-13, 235
Rhoads, Dr. Edward G., 217
Rhoads, Esther B., 217, 219, 221, 222, 223, 224, 225, 226-232, *passim*
Rhoads, Hannah, 258, 259
Rhoads, Dr. James E., 129, 130
Rhoads, Jonathan P., 216
Rhoads, Margaret W., 217, 222
Richards, Leighton, 144, 234
Richmond Conference (1887), 130
Richie, David S., 243
Ridgway, Jacob, 60
Roberts, Charles, 124
Roberts, Emmor, 104, 259
Robson, Elizabeth, 75, 184, 186
Rotch, Lydia, 182
Roudabush, Lillie, 228
Rushmore, Jane P., 6, 108, 197, 198, 256, 260
Russell, Elbert, 141, 147, 164

Salem General Meeting, 15, 21
Salem Monthly Meeting, 15
Salem Quarterly Meeting, 16, 34, 107, See also Haddonfield and Salem
Scarborough, John, 44
Scattergood, Alfred G., 235

Scattergood, Joseph, 104, 258
Scattergood, Joseph Henry, 212, 236
Schism, see Hicksite-Orthodox Separation
Schofield, Martha, 117, 194
Schwenckfelders, 203
Sellers, John, 61
Separation of 1827, see Hicksite-Orthodox Separation
Setagaya Neighborhood Center, 230
Sharpless, Amy E., 243
Sharpless, Edith, 217, 219-20, 222, 227, 230
Sharpless, Isaac, 111, 124, 131, 135, 147, 219, 235
Sharpless, Susanna F., 259
Sheppard, Clarkson, 258
Shillitoe, Thomas, 81
Shimazaki, Orie, 220
Shimotsuma, 219, 220, 230
Shrewsbury Monthly Meeting, 14
Shrewsbury Quarterly Meeting, 16, 42
Silcock, Harry, 225, 226-27
Simmons, Henry Jr., 206
Slave Trade, 44, 52-53, 91, 249, 251
Slavery, 18, 21, 22, 30-32, 44-45, 52, 63, 91-96, 182, 187, 192, 250, 251
Slavery, Germantown Protest against, 249
Smedley, Elizabeth, 259
Smiley, Albert K., 124, 212
Smith, Ephraim, 258
Smith, Esther Morton, 157-58, 190
Smith, John, 30
Smith, Samuel, 38
Smith, Sarah A.G., 222
Social Order Committees, 149, 150-52, 154, 242, 255, 257
Social Service Committee, 256
Some Considerations on the Keeping of Negroes, 31
South Africa, 246
Spanish-American War, 125, 141

Index

Special Peace Committee, (1916), 234
Stanton, Daniel, 38, 44
Stevenson, Ann, 258
Stockdale, William, 19
Story, John, 174
Strawbridge, Justus C., 133
Student Christian Union, 216
Student-in-Industry Program, 243
Sufferings, Meetings for, 39, 41, 45, 46, 47, 49, 52, 62, 69–70, 74, 75, 91, 92, 103, 104, 109, 113, 204, 205, 250, 254. See also: Representative Meeting
Sullivan, John A., 247
Swain, Joseph, 129, 139
Swaine, Joel, 206
Swan, Sarah C., 231, 232, 233
Swarthmore College, 61, 104, 111, 119, 128–29, 130, 135, 141, 164, 192, 195, 241–42, 253
Syria, 215, 220

Taxation for war, 40, 48, 49
Taylor, Francis R., 111, 218
Taylor, Howard and May, 229–30
Taylor, Jacob, 207
Taylor, Joseph W., 129
Taylor, Mary, 258
Temperance, 62, 96–99, 100, 106, 119–21, 156–57. See also Liquor
Temperance Association, 256
Test Oath, 41
"Testimony of the Society of Friends," 79, 252
Third Haven Monthly Meeting, 59
Thomas, Helen, 222
Thomas, M. Carey, 129
Tierney, Agnes L., 150, 190
Tobacco, 27
Tomiyama, Toki, 221, 228
Toyama Heights Meeting, 228
Toyama Neighborhood Center, 228, 230
Tract Society, 63, 83, 90. See also Friends Tract Association

Traveling ministers, 17, 26, 84, 110
Tsuchiura, 221
Tunesassa, 184, 185, 187, 206–07
Twelfth Street Meeting House, 237
Twelfth Street Monthly Meeting, 83, 118, 127
Two Weeks Meeting, 174

Ukaji, Kiyoshi, 228, 230
Underground Railway, 95
Unemployment, 154
"United meetings," 165
United Society of Friends Women, 186
Upland, 15, 248

Vaux, Roberts, 61, 88
Vietnam War, 16
Vining, Elizabeth Gray, 7, 227

Walker, James F., 169, 171, 257, 259, 260
Walking Purchase, 203, 250
Wallen, Sylvan, 232
Waln, Nicholas, 258
Walton, Anne Garrett, 110, 235
Walton, George A., 108, 141, 163, 168, 169, 260
Walton, J. Bernard, 108, 163, 235
Walton, Joseph, 103, 258
Walton, Joseph S., 104, 108
Walton, Margaretta, 108, 260
Waples, Douglas, 236
War taxes, 115–116, 155–56
Warder, Jeremiah, 45
Waring, Bernard G., 150
Warner, Hannah, 259
Warner, Yardley, 117
Warrington Monthly Meeting, 59
Washington, George, 49, 53, 251
Watanabe, Masa and Takeshi, 232
Weekend workcamps, 152
Welcome, The, 248
Welsh, Herbert, 211
Welsh Friends, 23
West Jersey, 15, 248

Western Quarterly Meeting, 25, 85, 197-98
Westtown Monthly Meeting, 78
Westtown School, 78, 83, 86-88, 98, 107, 110, 112, 135, 158-59, 183, 186, 188, 221, 251, 256
Wetherill, Samuel, Jr., 50
Wharton, Deborah F., 191, 259
Wharton, Joseph, 60
Wheeler, Daniel, 186
White, Allen J., 260
White, Barclay, 209
White, Esther, 180
White, Josiah, 60
Whittier, John Greenleaf, 124, 252
Whittier Fellowship, 141
Wilbur, Henry W., 133
Wilbur, John, 82, 83
Wilburites, 5, 82-83, 90, 103, 111, 112, 118, 126, 127, 131, 132, 135, 186
Wilkinson, John, 175
Willet, Rachel, 182
William Penn Charter School, 19, 29, 57, 149
William Penn Lectures, 161, 163
Williams, Mary R., 259
Wilson, Isaac, 140
Wilson, Woodrow, 125, 255
Wing, Asa, 216
Wistar, Edward M., 131
Women Friends, Yearly Meeting of, 85, 173-199, 253, 256, 258, 259, 260, 261
Women's Medical College, 133
Women's Problems Group, 151, 159
Women's suffrage, 151, 197
Women's Yearly Meeting, see Women Friends, Yearly Meeting of
Wood, Hollingsworth, 235
Wood, James, 131
Wood, Richard R., 166
Woodbrooke, 133

Woolman, John, 3, 31, 36, 37, 38, 40, 44, 83, 96, 141, 213. *Journal of*, 65
Woolman School, 141, 255
Work camps, 243
World Committee, see Friends World Committee for Consultation
World Conference of Friends, see Friends World Conferences
World Council of Churches, 8
Worship and Ministry, see Ministry and Worship
Worth, Herbert, 133
Wyoming Valley, 204

Yearly Meeting structure, 15-17, 25
Young Friends Conference, 163, 243, 255
Young Friends Movement, 140, 157, 160, 161-64, 167, 216, 222, 243, 256
Young People's Interracial Fellowship, 242
Young Friends in Japan, "The Remnant," 225, 226-27
Young Friends Peace Conference, 255